Fundamental
Phenomena
in the
Material Sciences

Volume 1
Sintering and Plastic Deformation

Fundamental Phenomena in the Material Sciences

Volume 1
Sintering and Plastic Deformation

Proceedings of the First Symposium on Fundamental Phenomena in the Material Sciences

Held February 1, 1963, at Boston, Mass.

Edited by

L. J. Bonis

President
Ilikon Corporation
Natick, Massachusetts

and

H. H. Hausner

Adj. Prof., Polytechnic Institute of Brooklyn
and Consulting Engineer
New York, N. Y.

Springer Science+Business Media, LLC
1964

ISBN 978-1-4899-6183-9   ISBN 978-1-4899-6367-3 (eBook)
DOI 10.1007/978-1-4899-6367-3

Library of Congress Catalog Card No. 64-20752

©1964 Springer Science+Business Media New York
Originally published by Plenum Press in 1964.
Softcover repint of the hardcover 1st edition 1964

FOREWORD

In recent years scientists and engineers working in the fields of metallurgy, ceramics, and polymer chemistry have come to agree that there are many basic phenomena which are common to metals, ceramics, and plastic-type materials. They have also discovered that a specialist working with any one of these three materials can gain considerably from wider knowledge of the other two. The new field of Materials Science has thus been developed to promote the advancement of all three types of materials.

Ilikon Corporation, located in Natick, Massachusetts, active in metallurgy, ceramics, and polymer research, development, and production, and strongly involved in the interdisciplinary aspects of materials, organized a symposium on "Fundamental Problems in the Materials Sciences," held on Feb. 1, 1963, in Boston, at which experts in all three types of materials had an opportunity to discuss common problems and exchange ideas. The main topics of this symposium were sintering and plastic deformation of each of the three types of materials. Sintering, the most important step in converting powder materials into solid products, is applied in metallurgy and ceramics, as well as in polymer technology. Diffusion is usually regarded as the mechanism for the solid state transport of material during sintering. Nevertheless, it appears doubtful that all sintering phenomena can be attributed to this mechanism. Recent work, described in the chapters of this book, indicates that plastic flow also must be an important mechanism in material transport during sintering.

The 1963 Boston symposium, on which the chapters of this book are based, attracted such a large audience that the meeting room had to be changed at the last minute in order to make space for all the participants. The wide participation and high level of the discussions that followed the presentation of the papers gave evidence of the need to bring scientists and engineers from the three related material fields together. The papers were presented by experts from Rensselaer Polytechnic Institute, Massachusetts Institute of Technology, Brown University, and DuPont Experimental Station.

During the symposium, the majority of the participants indicated great interest in continuing this type of symposium on an annual basis; they also indicated their desire to have the six papers presented published, in order to have these presentations on record.

This volume contains the six papers presented at the symposium—three chapters on sintering and three chapters on plastic deformation—and the reader will become acquainted with the many similarities in the fundamental phenomena in metals, ceramics, and plastic-type materials, although the approach to these phenomena may not always be the same for these materials. It is believed that this book will be valuable for scientists, such as solid state physicists, chemists, engineers, and students of materials science, and will be helpful in stimulating new ideas for the development of materials in general.

Laszlo J. Bonis   Henry H. Hausner
President   Polytechnic Institute of Brooklyn
Ilikon Corporation   Consulting Engineer

ACKNOWLEDGMENT

The editors express their indebtedness to the individual authors for their cooperation in preparing the manuscript, to the Co-Chairmen, Professor F. V. Lenel and Professor G. S. Ansell, for their active interest in the subject, and last but not least to Mr. B. L. Mulhern for his help both in collecting the papers and in preparing the manuscript.

CONTENTS

Sintering

Plastic Deformation

Bibliography and Index

Sintering

Sintering of Metal Powders

F. V. Lenel

Rensselaer Polytechnic Institute, Troy, New York

INTRODUCTION

In this first session we consider the material transport in sintering. This is an application of physical chemistry and solid state physics to a very practical problem. Before I start a discussion of the physico-chemical and physical phenomena involved, I should like to make clear why this is a practical problem for the metallurgist, the ceramist, and the polymer scientist.

The problem is to produce a solid body from a material that is available in particulate form by a process involving elevated temperatures. All of you are aware that this is the fundamental problem in ceramic technology, e.g., to produce a brick from particles of clay, to produce a ferrite core from particles of iron and other oxides, and to produce a nuclear fuel element from particles of uranium dioxide.

The problem is fundamental again in one branch of metallurgy, the one appropriately enough called powder metallurgy. One of the oldest industrial applications of powder metallurgy was to produce from tungsten powder solid bodies of tungsten, which then could be fabricated into tungsten wire for lamp filaments. Another more recent application is in producing structural parts, such as gears and cams from iron powder. Producing solid bodies from powders is, of course, a specialized technique for metals compared with the more common technique of casting molten metal into a mold to produce a casting or, alternatively, to produce an ingot which is then further fabricated. This is not the place to discuss why certain metal products are fabricated from powder. Suffice it to say that an extensive technology, reaching into nuclear and aerospace applications, is based on powder metallurgy.

Now with respect to polymeric materials, it was extremely interesting for me to study the work of Dr. Lontz, some of which he will present later. The steps involved in processing Teflon powder into solid bodies of Teflon and the controls necessary in these steps are exactly parallel to those in processing, let us say, iron powder into solid bodies of iron.

3

Cold-pressing of Teflon powder, what the polymer chemist calls "pre-forming," corresponds to "compacting" of the powder metallurgist; the elevated-temperature treatment and the sintering step are parallel too. The problem of producing a solid body from particulate material by elevated-temperature processing is also an intensely practical problem for the polymer scientist.

In the example of the Teflon and the iron powder—and very similar examples could be mentioned in ceramic processing—the powders are first pressed at room temperature into compacts and are then sintered. The sintering is therefore done on a pressed compact. It is also possible to sinter an unpressed aggregate of loose powder. This may be, for instance, a slip casting. The physicochemical phenomena will be quite similar and, in some cases, somewhat simpler than in sintering compacts.

Occasionally in powder-metallurgical technology, and quite often in polymer technology, pressing and sintering may be combined into hot-pressing or sintering under pressure. In general, this method of processing, in which an exterior stress is applied, will not be discussed this morning, although I will have to say something about the boundary line between conventional sintering and sintering under pressure in my discussion of the sintering of metal powders.

One last boundary line must be drawn before I start discussing the fundamentals of sintering. We will confine ourselves to the sintering of material all of which has the same state of aggregation. In the case of sintering of metal powders this means a metallic crystalline solid. In the case of ceramics it may again mean a crystalline solid, this time usually an ionic compound, or it may mean a highly viscous amorphous material, in other words a glass. In the case of polymers it usually means such an amorphous glass. We are excluding sintering in which both a crystalline and a liquid phase are involved, not because this type of sintering is unimportant in either powder metallurgy or ceramic technology, but because its discussion would lead us too far afield.

SINTERING PHENOMENA

We now consider the phenomena observed in sintering, i.e., the geometrical changes which take place when a compact or a loose powder aggregate is heated at the sintering temperature for a given length of time.

What we observe are two phenomena, occurring simultaneously:

1. The contact area, often called the neck, between particles touching each other grows. At the same time the pores become rounded.
2. The loose powder aggregate or compact becomes denser, i.e., the total pore volume decreases and the average distance between the centers of particles decreases

The geometrical changes occurring during the sintering of metal powder have been studied: (1) in model experiments, examples of which are a sphere on a plate, wire wound on a rod or a spool of wire on a rod, and three wires twisted together; and (2) in actual loose powder aggregates or compacts in which were observed the changes in pore shape, pore size and size distribution, and pore volume.

There has been a tendency to generalize the results of model experiments and to assume that a quantitative explanation of a model experiment also means a quantitative explanation of all sintering phenomena. This tendency to generalize must be guarded against. On the other hand, without the model experiments our understanding would not have advanced to its present stage.

To understand the nature of the changes in geometry, we ask two questions:

1. What are the driving forces?
2. What is the mechanism of atom movement?

The most important driving forces in sintering are surface tension forces. They exist in crystalline solids, as well as in liquids or in glasses. These forces tend to (1) increase the radius of curvature in an irregular pore, until an equilibrium configuration is obtained which takes into account the surface tension of the free surfaces and the interfacial tension forces at grain boundaries between grains or particles with different orientation; and (2) decrease the total surface area of the pore by shrinking the pores.

We see that the action of the surface tension forces causes increase in neck area, rounding and spheroidization of the pores, and shrinkage of the pores and densification.

It is generally agreed that surface tension forces cause only shrinkage of the compacts, not expansion. In compacts of single metal powders any expansion or growth is attributed to the action of gases, entrapped in the pores of the compact during compacting and initial densification, or to desorption of gases etc. The effect of the gases will be quite different depending upon whether:

1. They dissolve in the metal and diffuse through its lattice, e.g., hydrogen in copper.
2. They react with the metal, forming a solid compound, e.g., oxygen in copper.
3. They are insoluble and do not react, e.g., nitrogen in copper.

Much of the dimensional-change behavior in metal powder compacts, e.g., shrinkage followed by expansion during isothermal sintering of copper powder compacts can be understood, at least in a qualitative way, by considering the balance of forces of contraction, i.e., surface tension forces, which will depend upon the average diameter of the pores and forces of expansion due to the pressure of the gas in closed-off pores which will increase with increasing temperature, but will be independent of the diameter of the pores.

It is interesting to note that effects of gases in the sintering of ceramic compacts, in particular the dependence of the final density of compacts upon the atmosphere in which they are sintered, can be explained on exactly the same basis as the effect of gases in metal powder compacts, i.e., diffusion or lack of diffusion of the gases trapped in pores.

Surface tension forces are the most important driving forces, but not the only forces which cause changes in pore geometry of metal powder aggregates or compacts during sintering. A recent investigation has shown that gravity forces also contribute to the dimensional changes, in the sintering of both loose powder aggregates and compacts. In loose powder aggregates shrinkage in the vertical direction is greater than in the horizontal direction, regardless of the manner in which the loose powder is filled into the mold. In powder compacts the radial shrinkage of the compact will be different at the top and at the bottom section of the compact. In compacts which are sintered conventionally, i.e., with the bottom of the compact resting on a support in the sintering boat, the radial shrinkage near the top of the compact will be several percent larger than near the bottom of the compact, upon which the entire weight of the compact rests. On the other hand, in compacts which are suspended from the top, the radial shrinkage will be greatest near the bottom of the compact.

These observations indicate that in terms of driving force there cannot be any sharp distinction between conventional sintering and hot-pressing. The stress due to gravity which must be taken into account in conventional sintering is of the same nature as the externally applied stress in hot-pressing. With the same type of driving force operating in conventional sintering and hot-pressing there is no reason to make a distinction in the mechanism of material transport for the two methods of sintering, unless the magnitude of the stress applied in hot-pressing is much greater than the stress due to surface tension forces.

Another possible force for dimensional changes in compacts is the residual stresses which are introduced in compacts during pressing. When the dimensional changes during sintering are followed with a dilatometer, the first shrinkage in compacts is observed at temperatures at which these residual stresses are not yet relieved. In copper powder compacts pressed at 10,000 psi, the beginning of shrinkage is observed near 200°C, while the residual stresses will not be relieved until the compacts are heated to temperatures above 400°C. By correlating the magnitude of the stresses with the temperature of beginning shrinkage it can be shown that at least in the low-temperature range, up to 400°C in copper, residual stresses and externally applied stresses cause shrinkage. It should be emphasized that the primary causes of shrinkage during commercial sintering are undoubtedly the surface tension forces. Other forces contribute, but are not the most important ones.

We should now take a look at the mechanism of material transport which causes an increase in neck area and shrinkage in metal powder aggregates and compacts. In contrast to the studies of the mechanism of material transport in ceramics, where much of the experimental work was done on actual compacts, most of our knowledge of the material transport mechanism in the sintering of metals stems from model experiments rather than experiments on actual loose powder aggregates or compacts.

At least in the case of compacts it is not so surprising that the interpretation of experimental results for metal powders is more difficult than for ceramic powders. When metal powders are pressed, it is inevitable that the metal is strain-hardened. During sintering the effects of relief of strain hardening are superimposed upon those of material transport under the influence of surface tension forces, which makes the interpretation of the results difficult. This difficulty does not exist in ceramic powders, where little strain hardening occurs during pressing.

The first, and by now classical, model experiments on the mechanism of material transport in sintering powders are those of Kuczynski, first published in 1949[1]. These are the experiments in which Kuczynski observed the increase in the neck area between a spherical powder particle resting on a flat plate with increasing sintering time. He found that the relation between this radius x and the sintering time t was

For glass powder spheres

$$x^2 \propto t$$

and

For metal powder spheres

$$x^5 \propto t$$

On the basis of a calculation which involves certain simplifications, Kuczynski was able to show that the relationship $x^2 \propto t$ for glass powder spheres indicates a viscous creep material transport mechanism in the amorphous glass spheres and the $x^5 \propto t$ relationship indicates a vacancy diffusion mechanism of material transport for the metal powder spheres.

Kuczynski answered the question as to where in his model experiments the lattice vacancy gradient necessary for diffusion originates by pointing out that the volume of a solid underneath a surface which has a very small radius of curvature has a higher equilibrium vacancy concentration than a volume of a solid underneath a flat surface. Therefore a vacancy flux is established from the edge of the neck to the flat surface or to the adjacent concave surface of the sphere. Such a vacancy diffusion flux would explain quite satisfactorily the growth in contact area between a sphere and a flat or between two spheres.

It is difficult, however, to see how such a vacancy flux can satisfactorily explain the observed volume shrinkage of metal powder aggregates. To explain such shrinkage another type of sink for vacancies is necessary. This type of vacancy sink is provided by the grain boundaries in a theory postulated independently by Nabarro[2] and by Herring[3]. According to this theory a vacancy flux is established in a polycrystalline material subjected to a stress from grain boundaries under a normal tensile stress toward boundaries where there is a compressive stress. The boundaries under tensile stress in the theory would be the free surfaces, which are subject to surface tension, while the vacancy sinks would be provided by grain boundaries near the free surfaces.

The operation of this Herring–Nabarro flow has been clearly demonstrated in experiments by Balluffi and Alexander[4] who sintered artificial compacts consisting of layers of wire wound in the form of a spool upon a rod. Balluffi and Alexander were able to show first the spheroidization of the cross section of the cylindrical pores in their compacts and then the decrease in the diameter of the cylindrical pores until finally the grain boundaries between the pores disappeared through grain growth, which caused the pore shrinkage to practically cease.

Additional experiments supporting this theory were made by Kuczynski[5]. He used models consisting of three wires twisted together and measured, on the one hand, the growth in contact area, and, on the other hand, the change in the distance between the centers of the wires, which is a measure of densification. Since Kuczynski was able to produce three wire models with and without a grain boundary along the contact area of the wires, he could demonstrate quantitatively the role of grain boundaries in the sintering of his models.

The question as to what happens to the vacancies once they have diffused to the grain boundaries has been asked repeatedly. Brett and Seigle[6] showed experimentally that the vacancies are not eliminated by rapid grain-boundary diffusion to the outside surface of the samples. A theoretical model based on the operation of a dislocation mechanism to eliminate the vacancies has recently been advanced by Van Bueren and Hornstra[7].

It has been shown that the geometrical changes in model experiments on wire spools and twisted wires of copper and nickel at temperatures near the melting point of the metals follow quite well what would be expected from the Nabarro–Herring diffusion mechanism of material transport. Nevertheless, it appears doubtful that all sintering phenomena can be attributed to this mechanism. As will be brought out this afternoon the most important mechanism of deformation in metals is plastic flow by the movement of dislocations. Whether plastic flow by dislocation movement can contribute to material transport in the sintering of metals has been hotly argued. Recently, we have begun a series of experiments at Rensselaer Polytechnic Institute which seem to lead to the conclusion that plastic flow may be involved.

The first of these experiments were made on loose aggregates of spherical silver powder which were put under a small external stress by placing a silver electrode on top of the aggregate. The changes in electrical conductivity of these aggregates with time were measured at several temperatures in the range near 300°C. These changes in electrical resistance were interpreted as changes in contact area between the particles. By measuring the rate of change of contact area at several temperatures the activation energy of the process could be derived. It was found that this activation energy is not constant, but depends upon the external stress applied. Parallel experiments were made on the stress dependence of the activation energy for creep in silver in the same temperature range. The stress dependence of the activation energy of both processes is the same and can best be interpreted as a cross-slip mechanism which governs both increase in contact area and creep.

The temperature at which these experiments were made is below half of the melting temperature of silver in degrees absolute. One would therefore expect little volume diffusion to take place. At higher temperatures rates of diffusion become appreciable. Diffusion controls the Nabarro–Herring transport mechanism, but it also plays an important part in determining the rate of at least one mechanism of plastic flow by dislocation movement, i.e., the rate of steady-state creep, where this rate is governed by the recovery mechanism of dislocation climb. Even though a plastic flow mechanism controlled by both dislocation climb and the Herring–Nabarro transport mechanism would be expected to exhibit the same activation energy, i.e., the one of self-diffusion, the rates of the two processes would be quite different. Critical experiments to decide between these two mechanisms are not easy to find. One of the distinguishing features of the processes is their stress dependence, but it is difficult to express the action of the surface tension forces in the form of stresses.

A qualitative experiment which we recently made seems to point toward creep by dislocation climb as the controlling mechanism in sintering compacts made of two types of spherical copper powder at approximately 1000°C. One of these two types of powder consists of pure atomized copper; the other was made from a copper-$\frac{1}{4}$ w/o aluminum alloy powder which was internally oxidized so as to produce a fine dispersion of aluminum oxide in pure copper. For both types of compacts the rate of shrinkage during sintering at constant temperature was determined at a series of temperatures. As expected the heat of activation of both processes was quite close to the heat of activation of self-diffusion in copper.

However, the rates of shrinkage were considerably lower in the internally oxidized powder compacts than in the pure copper powder compacts. One would expect that the vacancy diffusion mechanism of Nabarro–Herring would not be slowed up by the presence of second-phase particles, which, on the contrary, may act as sources and sinks of vacancies. The dislocation climb mechanism, on the other hand, should be slowed up considerably by second-phase particles, which make the climb of dislocation around and over them more difficult, as was shown by Ansell and Weertman[8].

What role in the sintering of metals is played by a plastic deformation process such as steady-state creep governed by dislocation climb is not yet clear and a more quantitative approach will be necessary. The question as to whether plastic deformation should be considered in the sintering of ceramics, has not even been asked, which goes to show that our understanding of sintering is by no means complete.

REFERENCES

1. G. C. Kuczynski, *Trans. AIME* 185:169 (Feb. 1949).
2. F. R. N. Nabarro, Report of a Conference on the Strength of Solids, The Physical Society, London (1948), p. 75.
3. C. Herring, *J. Appl. Phys.* 21:301,437 (1950).
4. B. H. Alexander and R. W. Balluffi, *Acta Met.* 5:666 (Nov. 1957).
5. H. Ichinose and G. C. Kuczynski, *Acta Met.* 10:209 (1962).
6. J. Brett and L. Seigle, 6th Annual Progress Report, SEP 259, Contract AT(30-1) 2102, General Telephone and Electronics Lab., Bayside, L.I., 1961.
7. H. G. Van Bueren and J. Hornstra, *Proceedings of the 4th International Symposium on the Reactivity of Solids,* Amsterdam, 1960, p. 112; J. Hornstra, *Physica* 27:342 (1961).
8. G. S. Ansell and J. Weertman, *Trans. Met. Soc. AIME* 215:838 (1959).

Ceramic and Metal Sintering: Mechanisms of Material Transport and Density-Limiting Characteristics

R. L. Coble

Ceramics Division, Department of Metallurgy, Massachusetts Institute of Technology, Cambridge, Massachusetts

INTRODUCTION

Professor Lenel's paper [1] covered the material on the derivation of sintering models [2,3], the features which distinguish different mechanisms of material transport, and the behavior of some metallic systems where the applicability of specific models has been demonstrated. It has been shown that a diffusion mechanism is generally operative in metal sintering. In this paper further data are presented showing that diffusion sintering occurs in ceramics. The main purpose is to discuss the similarities and differences in the behaviors of metals and ceramics considering solid-phase sintering, liquid-phase sintering, boundary diffusion effects observed in several systems, and some recent results on the mechanisms of material transport during hot-pressing. These are the various mixtures of systems which are sintered to fabricate materials of controlled density and grain size. We wish to understand what controls the density limit, what controls grain growth, and how these are interrelated. The interrelations are understandable if the mechanisms of matter transport that govern the kinetics of density change and of grain growth are known; this provides the reason for the emphasis on the mechanism of material transport in sintering studies.

SINGLE-PHASE SYSTEMS

In sintering single-phase crystalline systems, the behavior of typical ceramics, such as Al_2O_3, BeO, and MgO, may be compared with the behavior of typical metals, such as copper, iron, platinum, gold, silver, etc. The similarities in their sintering behaviors are several: (1) all can be sintered to densities in the range of 95% of theoretical without difficulty [2]; (2) all appear to undergo densification with material transport

11

by lattice diffusion [2,3]; (3) the sintering kinetics of all are affected by normal grain growth [2]; (4) the occurrence of discontinuous grain growth governs the maximum achievable density in most cases [4]; (5) gas trapped within closed pores at the final stage of sintering can control the density limit [5], or even cause density decrease with overfiring. The differences between metals and ceramics arise from their different deformability in powder pressing and their different response to oxidizing and reducing atmospheres during heating [1].

LIQUID-PHASE SINTERING

Tungsten carbide with cobalt may be selected as a typical metal system; for ceramics the porcelains may be selected for comparison [5,6]. When substantial amounts (less than 30% vol) of liquid phase form during the sintering process, densification or pore removal is governed by the viscosity of the glass or liquid phase [6] because there is little need for solution and reprecipitation of the crystal phase. With 30 to 50% liquid formed, particle rearrangement is a sufficient criterion for having the materials undergo complete densification. For smaller liquid contents (as may be used in WC) grain shape change is required for complete densification, and the two possible rate-controlling mechanisms are solution rate at the crystal/liquid interface or diffusion rate in the liquid.

For both metal and ceramic systems containing large amounts of liquid phase during sintering, complete densification is generally achievable. Two exceptions occur: (1) for metal systems sintered in vacuum, the liquid-forming metal is frequently vaporized prior to complete densification; (2) for ceramics sintered in air the gas trapped generally prevents complete densification.

TWO-COMPONENT SYSTEMS

The term "activated sintering" refers to the enhanced sintering rates which occur with additions of second components in small amounts, approximately 0.01%. For metals, the nickel-activated sintering of tungsten has been reported[7], and for the ceramic analogy, activated sintering of beryllia occurs with additions of calcium oxide. The effect is more pronounced in the tungsten–nickel system; tungsten is normally sintered at temperatures in the range of 2700°C to get near theoretical density, whereas with a nickel addition of about 0.01% theoretical density is reached at 1400°C. With a calcium oxide addition to beryllia there is a reduction by several hundred degrees Centigrade in the temperature required to achieve high density.

The content of second component required to produce the effect is the amount required to form a monolayer of the eutectic compositions between the known phases that appear in the respective phase diagrams. The rate increases linearly up to the content at which a monolayer forms and is unaffected with larger additions. It should be noted that the amount required to form a monolayer is dependent on the particle size of the host material.

The interpretation of this effect is that the film which forms is a region in which diffusion coefficients for the host material are much higher than the diffusion coefficient of the host material in pure grain boundaries.

Professor Lenel referred to the diffusion model where a pair of spheres are sintered together with vacancies diffusing from the neck surface or to the grain boundaries; the layers of material removed at the grain boundaries between particles permit them to move together [8]. For activated sintering, the model is exactly analogous to the boundary diffusion controlled case [9] with different assigned diffusion coefficients and widths of the diffusion zone.

QUANTITATIVE COMPARISON OF MEASURED DIFFUSION COEFFICIENTS WITH SINTERING KINETICS

For the single-phase materials the interpretation which is most widely accepted is that lattice diffusion is the rate-controlling mechanism of sintering and that the atoms diffuse from the grain boundaries to adjacent pores or solid–vapor surface. For the interpretation of metal sintering there is no problem in deciding which diffusion coefficients are important. For ceramics, however, there are independent diffusion coefficients for the cations and anions, and the rules of kinetics dictate that the diffusion coefficients calculated from sintering data be compared with the diffusion coefficient of the slower diffusing species. In the following we will see that in some cases this interpretation is correct, while in other cases an alternate interpretation is required.

In Fig. 1 the apparent diffusion coefficients calculated from the sintering of Fe_2O_3 compacts [9] are compared with the measured diffusion coefficients for iron ions in Fe_2O_3. The data reveal an agreement with one of the diffusion sintering models and show that it is the diffusion coefficient of iron which controls the process. In Fe_2O_3 it has been shown that oxygen diffuses more rapidly than iron and consequently the control of the kinetics by the slower diffusing species as the rate-controlling step is satisfied. In Fig. 2 the diffusion coefficients calculated from sintering data of Al_2O_3 are compared with the measured diffusion coefficients for aluminum ions and for oxygen ions in polycrystalline and single-crystal samples [10]. In Al_2O_3 the agreement exists between the measured diffusion coefficients for aluminum, while these values are significantly higher than the diffusion coefficients for oxygen in single crystals. Measurements on oxygen diffusion in polycrystalline alumina [11] showed that

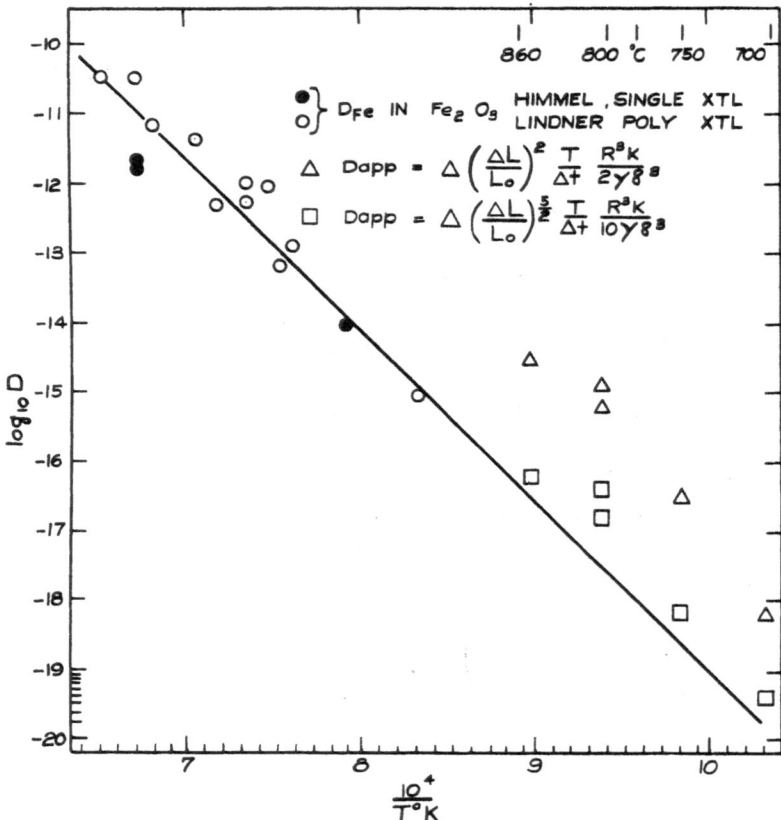

Fig. 1. Apparent diffusion coefficients calculated from shrinkage rates of Fe_2O_3 as a function of temperature compared with measured iron diffusion.

there is enhanced diffusion of oxygen ions at grain boundaries. We conclude that the rate of the process is controlled by the diffusion of aluminum through the lattice, while the necessary transport of oxygen takes place at grain boundaries [10]. This obviously constitutes an alteration of the interpretation which would have arisen through the simplest adaptation of classical kinetic models.

From this we conclude that for metals it is easy to select the appropriate diffusion coefficients in order to define the kinetics; it is not necessarily easy to select the appropriate diffusion coefficient for oxide sintering. The possible complication of enhanced diffusion at grain boundaries may require a modification of the model for the process.

The above data, where agreement between calculated diffusion co-efficients and directly measured values exists, are taken from measurements during the initial stages of sintering when essentially no grain growth occurs and the grain-size term which appears in the sintering expressions is constant. In the latter stages of sintering, grain growth occurs and the expression for the kinetics must be modified.

LATTER STAGES OF SINTERING: EFFECT OF GRAIN GROWTH

For the latter stages of densification of powder compacts, the data reveal a linear relation between density and log time [2]. The diffusion sintering model for the latter stage predicts that density changes linearly with time as given by equation (1),

$$\frac{\Delta\rho}{\Delta t} = \frac{10D\gamma\Omega}{(GS)^3 kT} \tag{1}$$

where D is the lattice diffusion coefficient, γ is the surface energy, Ω is the vacancy volume, GS is the grain size, kT is Boltzmann's constant times the absolute temperature, ρ is density, and t is time. For the derivation of this expression the grain size was assumed to be constant[2]. However, over a significant range of times in which density is observed to change, the grain size increases with the one-third power of time (Fig. 3)[2]. Therefore, to assess the densification rate an empirical grain size correction must be introduced. The resulting expression predicts that the density should change linearly with the logarithm of time, in accordance with what is roughly observed [2].

The diffusion coefficients calculated from the latter-stage sintering models are also presented in Fig. 2. The values are higher than measured diffusion coefficients for aluminum and the values calculated from the initial-stage sintering measurements. The support of latter-stage diffusion sintering models is based on the effects of grain size on rate and of the time dependence observed [2], rather than quantitative agreement between diffusion coefficients.

Measurements of the creep behavior of high-density alumina have shown that the Nabarro–Herring microcreep model applies and that the diffusion coefficients calculated from the creep rates are in good agreement with the measured aluminum ion diffusion coefficients [10]. Because a diffusional creep model in polycrystalline alumina applies over ranges of stress up to 20,000 psi and over a temperature range from 1000 to 1960°C, it seems reasonable to assume that in pressureless sintering a diffusional model must also apply. Therefore, from the data on diffusion coefficient, sintering, and diffusional creep, we conclude that sintering

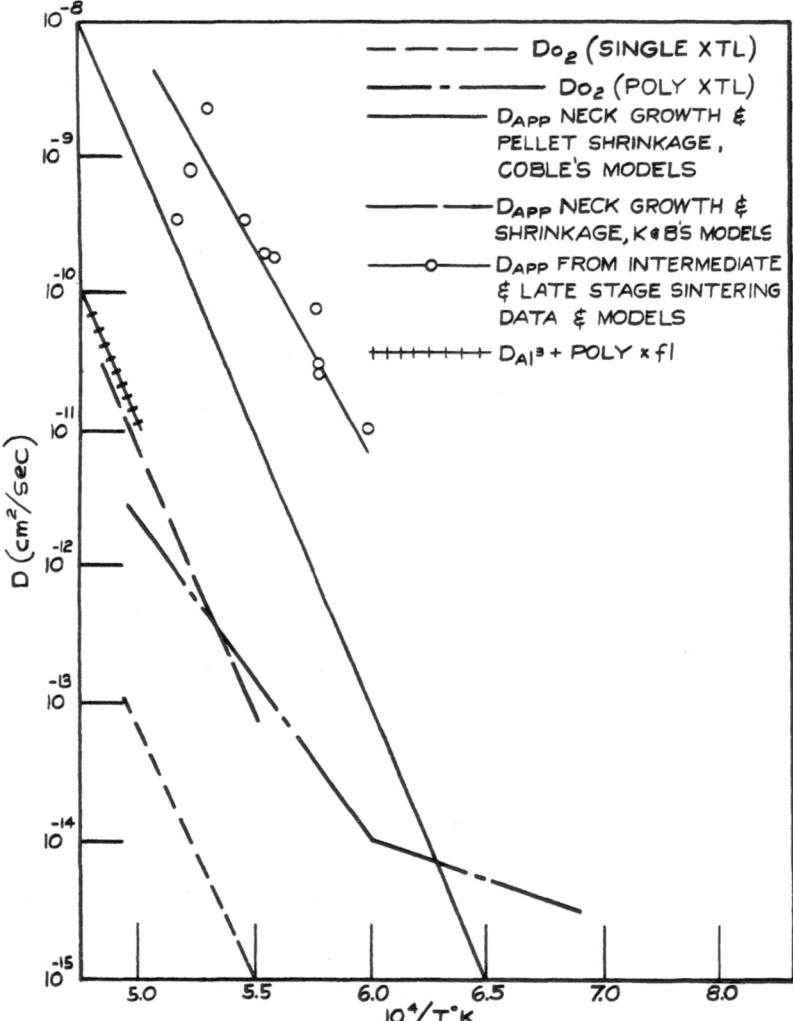

Fig. 2. Diffusion coefficients in alumina vs. reciprocal temperature directly measured aluminum and oxygen coefficients measured in single-crystal and polycrystalline samples [4] are compared with values calculated from sintering experiments and models.

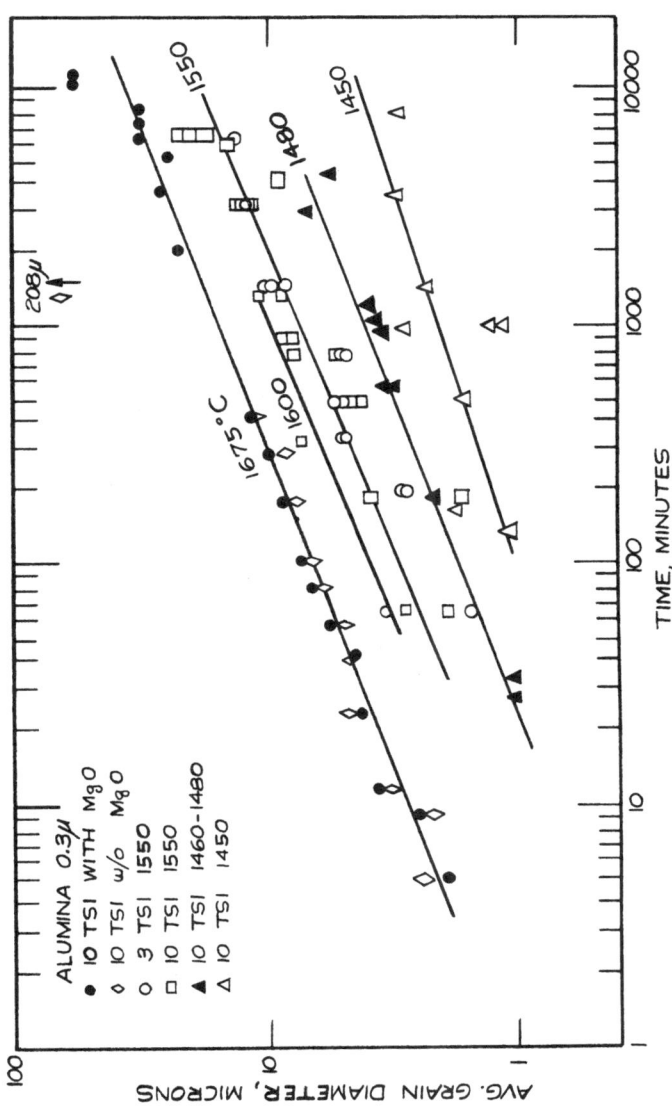

Fig. 3. Grain growth in alumina compacts with temperature, forming pressure and magnesia additive as variables.

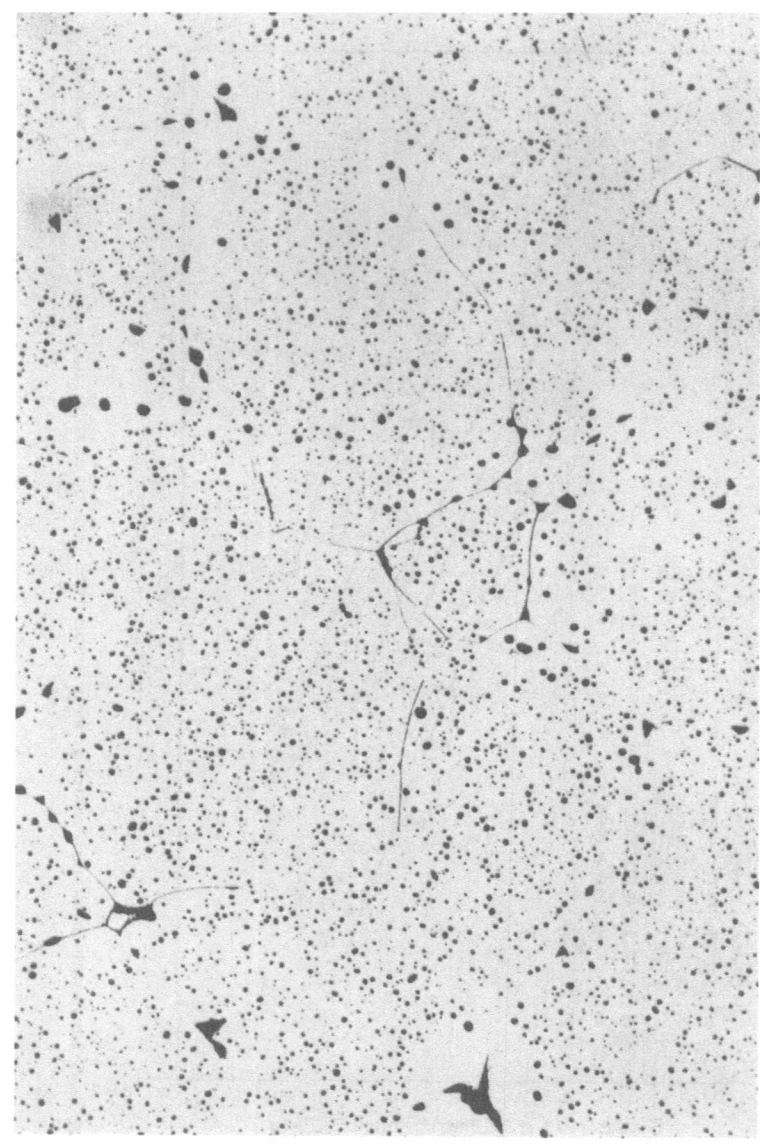

Fig. 4. Final stage after discontinuous grain growth. Photomicrograph at 100× α-alumina, 1950°C, 2 hr

does occur by diffusion and the numerical factors for the latter-stage sintering models are low.

DENSITY-LIMITING CHARACTERISTICS

The effects of normal grain growth on the kinetics of densification were shown above. In the majority of ceramics and probably in a number of metals, the occurrence of discontinuous grain growth governs the maximum density which can be achieved[4]. Figure 4 shows a microstructure of an aluminum oxide powder compact sintered to high density (97%). Discontinuous grain growth has occurred and trapped the pores inside the grains [3]. With prolonged heat treatment the porosity is eliminated from areas adjacent to grain boundaries.

The control of discontinuous grain growth in alumina permits the achievement of theoretical density [2]. For additions of second component at the solubility limit for growth control, the mechanism by which growth inhibition occurs is not fully understood. It is clear, however, that the densification rate must be increased relative to the grain growth rate. An alternate approach is to rely on particulate growth inhibition. By using a sufficient amount of second phase, discontinuous grain growth can be inhibited. If the occurrence of discontinuous growth arises at a critical density, the amount of second phase required for growth inhibition would be an amount in excess of the volume fraction pores at the critical density.

ATMOSPHERE EFFECTS

Professor Lenel pointed out [1] that the firing atmosphere may control the density limit. Compacts of silver and gold pressed to high pressures at room temperature exhibit density decrease upon heating. This is attributed to the fact that the pressure of air trapped inside the pores exceeds the equilibrium pressure governed by pore size and surface energy. Therefore, the system decreases in energy by allowing gas expansion and pore growth. Rhines et al[12] showed that the pore shrinkage occurs in copper compacts sintered in hydrogen, while pore growth occurs in argon. This result is consistent with the faster diffusion of hydrogen than argon in copper. Vines et al. [5] have shown how gas pressure affects the terminal density in porcelains. For aluminum oxide sintered in hydrogen, theoretical density is achievable. Alumina sintered in air equilibrates at a lower density as shown in Fig. 5. Equivalent low-density limits occur for alumina sintering in helium, argon, or nitrogen, while theoretical densities can be achieved in hydrogen, oxygen, or vacuum [13]. With respect to mechanical inhibition of full densification by gases trapped within pores, metal and ceramic systems are equivalent.

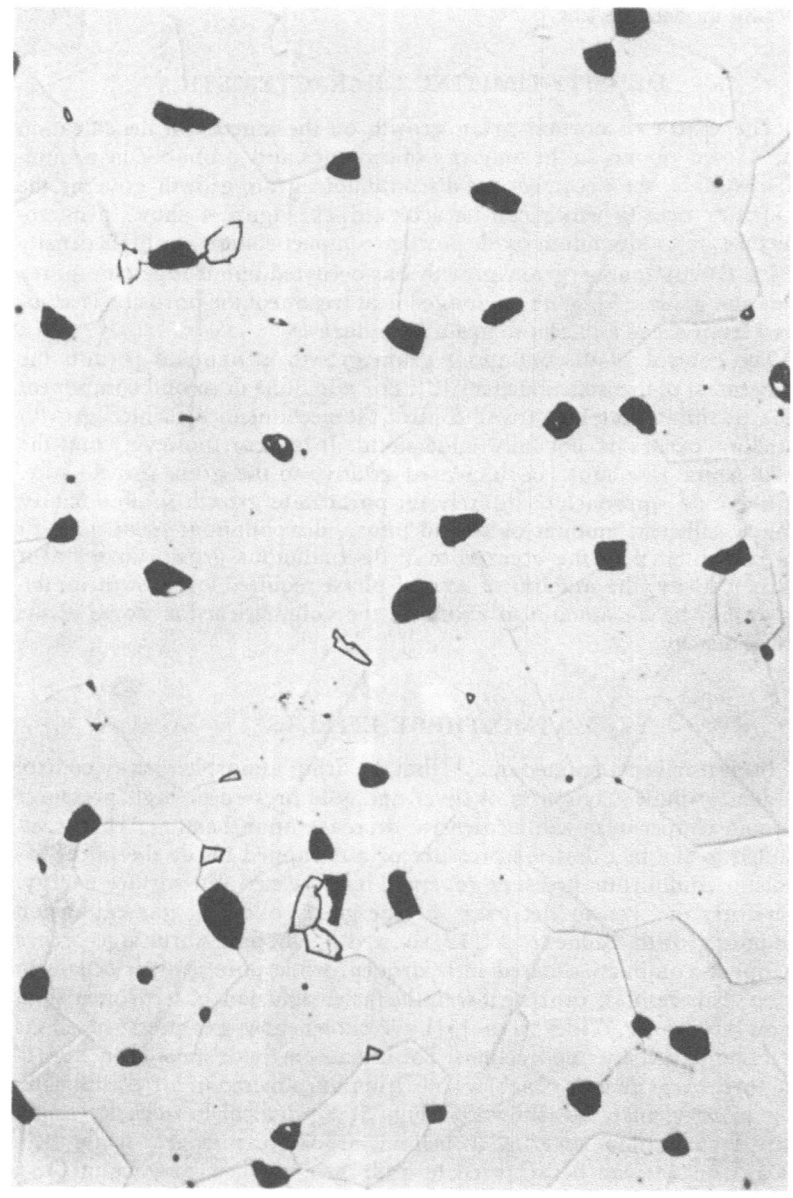

Fig. 5. Alumina-containing magnesia sintered in nitrogen for 72 hr at 1825°C.

HOT-PRESSING

A model system for studying hot-pressing has been set up [14] which is analogous to Kuczynski's [3] classical sintering experiments conducted between pairs of spheres. Single-crystal sapphire spheres, 40 mils in diameter, were pushed together under loads at several temperatures and times. The neck growth is determined from metallographic measurements of the contact areas after the spheres have been broken apart. The results are shown in Fig. 6 for 1500°C at three different loads. Also included in Fig. 6 are pressureless sintering data. The results show that under the influence of pressure the neck size is significantly enlarged, but within the experimental times (from 10 to 540 min) there is no time dependence for the process. The ratio of load to contact area formed is a stress which may be considered to be a self-hardness value. The value obtained for the measurements conducted at 1500°C is 50,000 psi; at 1700°C the stress is 30,000 psi; and at 1900°C it is 20,000 psi.

It was shown quantitatively [14] that the enhanced rate of net growth which occurred within 10 min cannot be attributed to Nabarro–Herring microcreep under the loads applied; the experimentally observed values are 100–10,000 times higher than the predicted values.

The equilibration of applied loads as self-hardness stresses provides the basis for assessment of the deformation which will occur upon loading particles in a hot-press die. Within a hot-press die, the total load applied to a particle may be assumed to be equal to the particle surface area times the applied pressure. The load is actually distributed among N contacts, where N is the average coordination number of the particles. At each contact point, deformation will occur until the stress is equilibrated at the self-hardness value with a given cross-sectional area such as occurs in the self-hardness measurement. The load balance is

$$\sigma_a \cdot 4\pi R^2 = N\sigma_{sh} \cdot \pi X^2$$

which may be rearranged to give

$$(X/R)^2 = 4\sigma_a/N\sigma_{sh} \qquad (2)$$

If it is assumed that the coordination number is 12 and that the applied pressure in hot-pressing is 10,000 psi, and σ_{sh} at 1500°C from measurements [14] is 50,000 psi, the calculated X/R value is 0.26. For a coordination number of 12, the initial density is 74% and the maximum X/R value for this coordination number is 0.6. Therefore, the actual densification which occurs by plastic flow is smaller than that required to complete densification during hot-pressing. We conclude that under these conditions the final densification of aluminum oxide probably occurs by a modified Nabarro–Herring microcreep (diffusional) mechanism.

By increasing the temperature of hot-pressing to 1900°C, the self-hardness value drops to 20,000 psi. For an applied pressure of 10,000

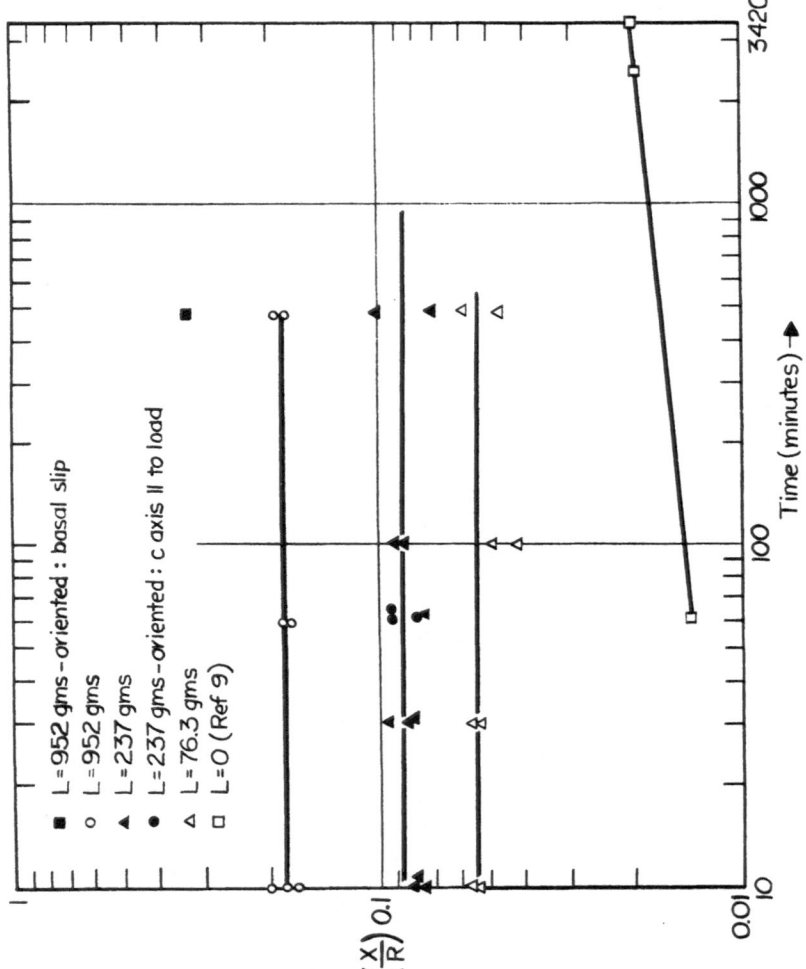

Fig. 6. Log-log plot of x/R values observed under loads indicated.

psi the X/R value is approximately 0.4. Thus, even at this higher temperature (and 10,000 psi) complete densification by plastic flow would not occur for aluminum oxide.

It seems probable that the final densification in materials like aluminum oxide will occur by pressure-directed diffusion rather than by plastic flow. It is possible that materials like magnesium oxide, sodium chloride, and other compounds with multiple slip systems will undergo densification by plastic flow.

The distinction can probably be made most easily by noting whether the creep process in a given material at full density at a stress and temperature at which hot-pressing is to be conducted occurs by a diffusion mechanism or by a plastic-flow mechanism. It is probable that the materials which undergo diffusional creep at full density will exhibit final densification in hot-pressing by the pressure-directed diffusional mechanism. This is the case with aluminum oxide. For materials which deform by plastic flow, it seems probable that their final densification will also occur by plastic flow.

Again, there is no reason to distinguish between ceramics and metals for their behavior in hot-pressing, with appropriate atmospheres for the respective classes of materials.

REFERENCES

1. F. V. Lenel, "*Sintering of Metal Powders,*" this volume, p. 3.
2. R. L. Coble, *J. Appl. Phys.* 32:789 (1961).
3. G. Kuczynski, *J. Appl. Phys.* 21:632 (1950).
4. J. E. Burke, *J. Am. Ceram. Soc.* 40:80 (1957).
5. R. F. Vines, J. O. Semmelman, P. W. Lee, and F. P. Fonvielle, *J. Am. Ceram. Soc.* 41:304 (1958).
6. W. D. Kingery, E. Niki, and M. D. Narasimhan, *J. Am. Ceram. Soc.* 44:29 (1961).
7. J. Brophy, *et al,* International Conference on Powder Metallurgy, 1960, New York.
8. W. D. Kingery and M. Berg, *J. Appl. Phys.* 26:1205 (1955).
9. R. L. Coble, *J. Am. Ceram. Soc.* 41:55 (1958).
10. A. E. Paladino and R. L. Coble, *J. Am. Ceram. Soc.* 46:133 (1963).
11. Y. Oishi and W. D. Kingery, *J. Chem. Phys.* 33:480 (1960).
12. F. N. Rhines, *et al, Trans. AIME,* 188:378 (1950).
13. R. L. Coble, *J. Am. Ceram. Soc.* 45:123 (1962).
14. R. L. Coble and J. Ellis, *J. Am. Ceram. Soc.* 46:438 (1963).

Sintering of Polymer Materials

John F. Lontz

E. I. du Pont de Nemours & Company, Inc., Plastics Department
Dupont Experimental Station, Wilmington, Delaware

INTRODUCTION

Polymers comprise an important segment of engineering materials with ever-increasing applications in structural components, supplementing and often replacing metals, ceramics, and wood products. As with metals and ceramics, the fabrication or conversion of polymers to useful forms involves fundamental aspects of structure and properties such as are being considered in this symposium In consolidating or shaping the polymers by melting or fusing, the rheological concepts of material transport involve motion or flow that may in some sense have a common basis with ceramics, particularly glass, and possibly powdered metals especially in technology of shaping useful objects. Although such commonality is limited, the techniques and tools for gaining an insight or understanding of how polymer consolidates to tough, durable materials can often be quite similar.

POLYMERS AS VISCOUS MATERIALS

Polymers can be regarded as being consolidated during melting or fusion through the action of forces that involve two principal rheological attributes, namely, the viscous and the elastic characteristics. These attributes determine the movement or transport mechanism under the action of stress, imposed either by direct mechanical action or by thermal conditions or by the combination of the two. Additionally, and to be more complete, surface tension plays a role, but only a minor one since the range of 25 to 50 dyn/cm is relatively insignificant compared to that involved with metals and ceramics.

The viscous and the elastic characteristics of a polymer during consolidation or, as in the case of creep, distortion are not mutually exclusive. In fact, the same polymer, depending upon existing conditions of stress and temperature, can assume the viscous and elastic extremes as well as an intermediate viscoelastic quality.

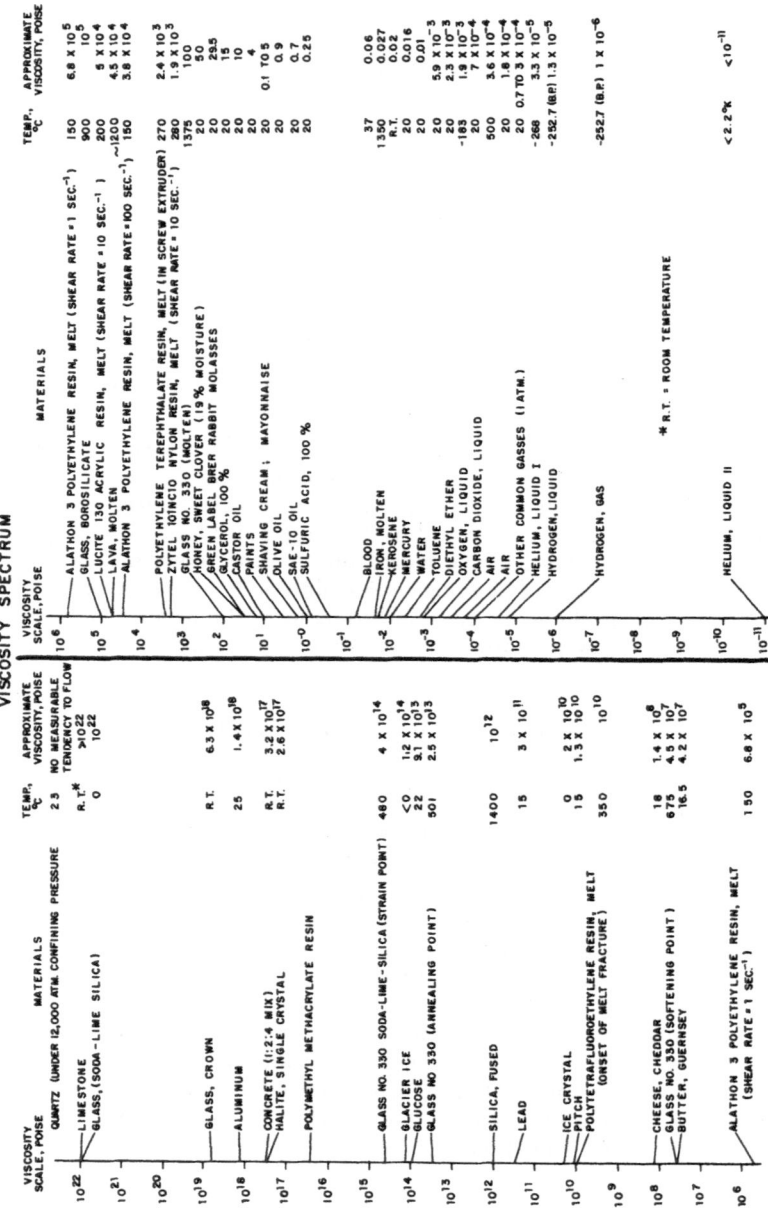

Fig. 1. Viscosity spectrum of diverse materials [1].

Viscosity Features

In general, all materials regardless of their physical state, gas, liquid, or solid, are endowed with a degree of viscous attribute such as is indicated in the viscosity spectrum of Fig. 1. Included in addition to polymers are diverse materials as a reference scale to indicate differences as well as similarities among various materials. The magnitude of the assigned viscosity value depends upon the intermolecular forces, which in turn are determined by molecular size, shape, polarity, and other physical features. For instance, in a given homologous series, the higher the member, the higher the molecular weight and the higher is the viscosity; indeed, viscosity measurements provide an important technique for determining molecular weights. As the homologous series extends to long molecular chains numbering several orders of magnitude or higher, the size of the moving elements undergoing viscous flow increases as do other flow properties, as shown in Table I [2]. In this table there are introduced such features as the average volume of moving elements, the activation energy or a measure of flow response to temperature, and a relaxation time representing a measure of restorative or inertial quality. All of these features are measurable variations of the basic viscous attributes, each having an important role in sintering or coalescence of polymer materials.

Elastic Characteristics

In any transport phenomenon including motion in viscous flow, numerous mechanisms are brought into play that involve elastic deformation down to the bond level. A general listing of the elastic range for various solids, including polymers, in terms of the descriptive mechanisms is summarized in Table II [2]. This tabulation is intended to distinguish the various levels of elastic deformation in terms of elastic modulus, which in turn is an indication of resistance or yield under applied physical stress. The mechanisms as listed emphasize that in any sintering or coalescence, where some degree of distortion is bound to take place under direct or indirect (thermal) stress, a whole spectrum of bending or stretching episodes can be involved. This would also mean that during fusion there could develop restraints from material components with elastic moduli ranging from 10^6 to 10^{12} dyn/cm^2. The restraints in effect retard the flow in the case of polymers by virtue of numerous structural characteristics that include chain entanglement, cross-linking (including hydrogen bonding), and various degrees of both diffuse and well-defined order (including crystallinity). By constructing appropriate models depicting the viscous and elastic attributes, it is possible to develop workable concepts of the corresponding retardation time or constants, as discussed in the ensuing section.

TABLE I

Flow Properties of Various Materials [2]

Moving element	Average volume, A^3	Average energy of activation, kcal per moving element	Viscosity at melting point, poises	Relaxation time, sec	Typical example
Single atoms	10	1	10^{-3}	10^{-13}	Argon
Single molecules					
Weak forces	50–500	2	10^{-2}	10^{-12}	Pentane
Medium forces		3–4	10^{-1}–1	10^{-11}–10^{-10}	Acetaldehyde
Strong forces		4–6	1–10	10^{-10}–10^{-9}	Octyl alcohol
Aggregates of molecules	10^2–10^4	10	10–10^2	10^{-9}–10^{-8}	Glycerol
Segments of macromolecules	10^2–10^4	20–40	10^3–10^5	10^{-7}–10^{-5}	Vinyl polymers
Aggregates of macromolecules	10^5–10^8	>100	>10^5	>10^{-5}	Cellulose esters

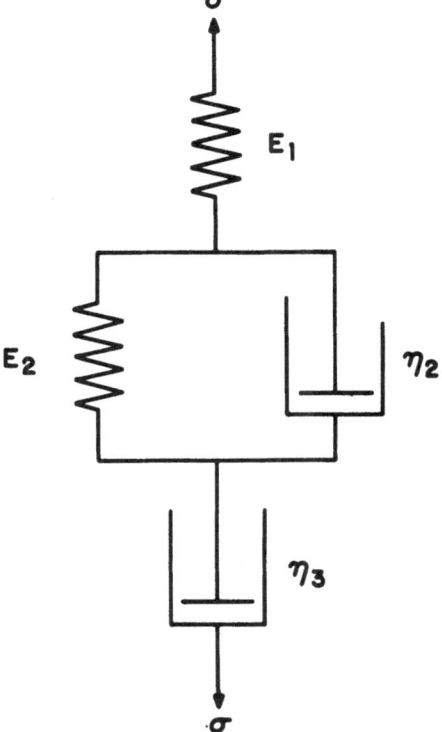

Fig. 2. A four-element Maxwell–Voight
model for viscoelastic flow [3].

Retardation Time

As the polymer undergoes deformation, for instance under tensile stress, it is convenient to visualize by means of the Maxwell–Voight four-element model shown in Fig. 2 [3] the role of the viscous and elastic attributes in a typical time-dependent creep curve shown in Fig. 3 [3]. The four-element model in Fig. 2 assumes that each of the two springs (E_1 and E_2) is Hookean in nature, while the dashpots (η_2 and η_3) correspond to pure Newtonian liquid. When the stress σ is applied, it is instantly taken up by spring E_1 and dashpot η_3, followed in time by the delayed spring E_2 and dashpot η_2 constants connected in parallel. As the stress continues with time, the elongation ϵ_2 of the parallel spring E_2 and dashpot is expressed as

$$\epsilon_2 = \frac{\sigma}{E_2}[1 - \exp(-t/\tau)] \tag{1}$$

TABLE II

Elastic Properties of Solid Materials [2]

Mechanisms involved in elastic deformation	Typical examples	Elastic modulus, dyn/cm^2	Range of elastic extensibility, %	Dependence upon temperature, T
Stretching of primary valence bonds	Ionic bonds: salts Metallic bonds: nickel Covalent bonds: diamond	$5 \cdot 10^{12}$ to $5 \cdot 10^{11}$	0.1 – 0.5	Slight – proportional to $1/T$
Bending of covalent primary bonds	Quartz, diamond	$5 \cdot 10^{11}$ to $5 \cdot 10^{10}$	0.2 – 0.8	Slight – proportional to $1/T$
Stretching of hydrogen bonds	Ice	$5 \cdot 10^{11}$ to $5 \cdot 10^{10}$	0.2 – 0.8	Slight – proportional to $1/T$
Stretching of van der Waals' bonds	Sucrose, phthalic anhydride, naphthalene	$10^{11}–10^{10}$	0.5 – 1.5	Slight – proportional to $1/T$
Stretching of secondary valence bonds combined with kinetic elasticity (uncoiling) of high-polymer chains	Strong forces: cellulose, nylon, proteins Medium forces: polyesters, polyvinyl alcohol Weak forces: hydrocarbon polymers	$10^9–10^8$ $10^8–10^7$ $10^7–10^6$	Up to 1000	Proportional to T

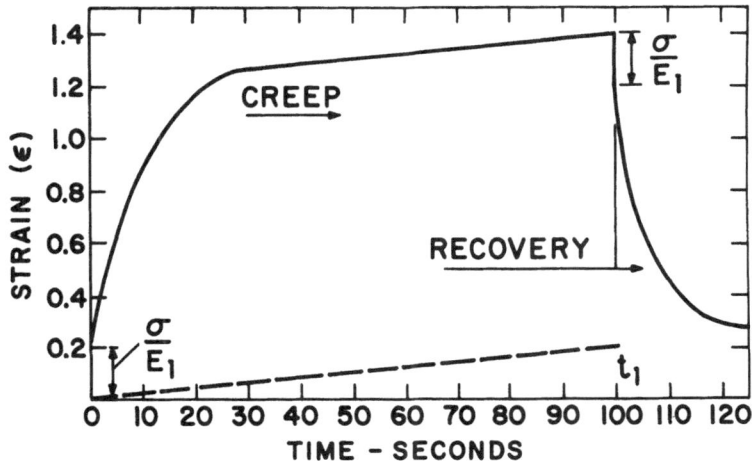

Fig. 3. A typical time-dependent creep [3].

where τ, the retardation, is defined by delayed constants as

$$\tau = \eta_2/E_2 \tag{2}$$

The time-dependent strain responding to an applied stress σ of 10^9 dyn/cm^2, as shown graphically in Fig. 3, is comprised of the following illustrative Hookean and Newtonian constants:

$$E_1 = 5 \times 10^9 \text{ dyn/cm}^2 \qquad \eta_2 = 5 \times 10^9 \text{ poises}$$
$$E_2 = 1 \times 10^9 \text{ dyn/cm}^2 \qquad \eta_3 = 5 \times 10^{11} \text{ poises}$$

and hence

$$\tau = \eta_2/E_2 \text{ or } 5 \text{ sec}$$

It is possible then to conceive of retardation times with several orders of τ values which would be inherent in or characteristic with the polymer structure. Hence, this constant serves as an important characteristic in defining the sintering quality, which in turn represents the time involved in delayed readjustments of stressed molecular configurations. Experimentally, stress-relaxation times [3] are determined from creep measurements in order to characterize given polymers as a means for assessing the validity of analogous assigned retardation constants from model considerations.

It is interesting to note, as shown graphically in Fig. 4, a correlation between relaxation time and viscosity, as listed earlier in Table I. The graphical plot is intended to indicate that with increasing viscosity, and hence through increasing molecular weight, polymers may acquire high

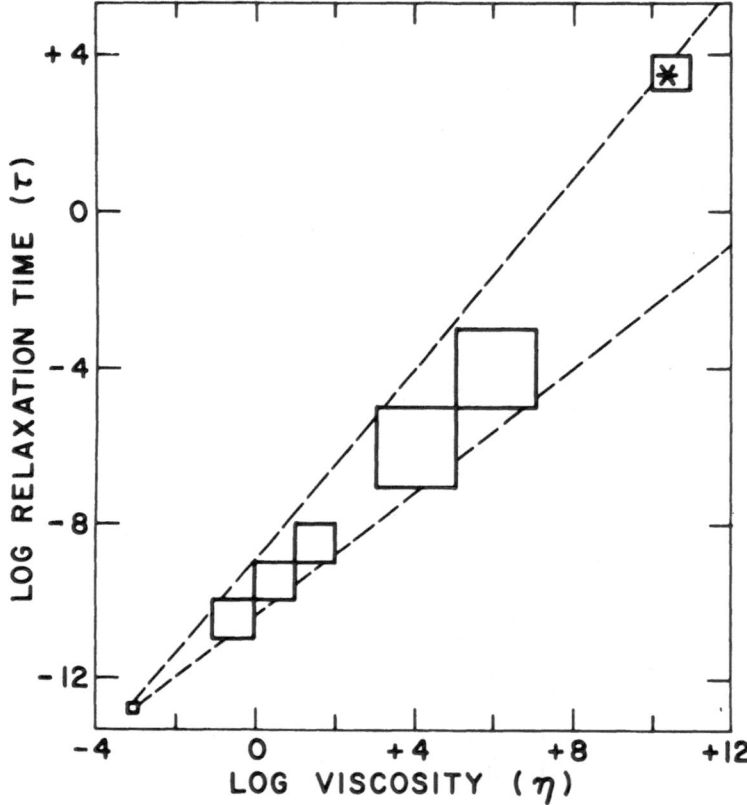

Fig. 4. Approximate correlation of relaxation times for viscosity.

relaxation constants, such as indicated for polytetrafluoroethylene [4]. The latter attains viscosity to the level of 10^{10} to 10^{11} poises [5], and hence acquires a significant retardation constant 10^4 sec, which demands measurably high sintering exposure times.

   With this more or less general description of the significance of the viscoelastic nature of polymers as an indispensable characteristic in material flow or transport, we can proceed to consider fusion or coalescence with two general cases. These would include first considering purely viscous flow as is generally applied in melt extrusion technology and to some degree in film-forming of lattices. This will be followed by the second case involving sintering viscoelastic polymers, notably polymethyl methacrylate and polytetrafluoroethylene.

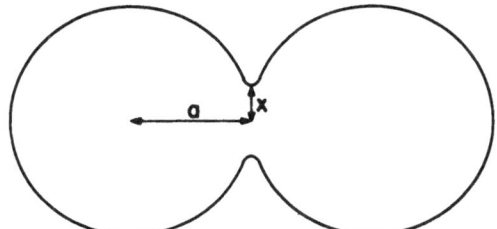

Fig. 5. Model for sintering concept. Two spheres of radius a to form a contact area of radius x.

VISCOUS SINTERING OR COALESCENCE

The concept of viscous sintering was first developed by Frenkel [6], who derived an expression for the rate of coalescence of adjacent spheres under the action of surface tension, the thermodynamic premise being that a system left to itself would lower its free energy by decreasing its total surface. From the model given in Fig. 5, the Frenkel expression takes the form

$$x^2 = 3a\gamma t/2\eta \tag{3}$$

where x is the radius of the interface between the two spheres, a is the radius of the sphere, γ is surface tension, t is time of contact, and η is the viscosity.

This relationship has been confirmed by Kuczynski, using glass at temperatures that provided a viscosity range from 10^6 to 10^8 poises, and accounted adequately for the first of two stages of sintering, namely, developing interfaces or bridges between adjacent particles with little change in density. For the second stage, normally described as densification, in which the more or less spherical cavities are eliminated, Kuczynski [7] was able to apply the glass sintering mechanism to the expression

$$r_0 - r = \gamma t/2\eta \tag{4}$$

where r_0 is the original radius of the pore and r is the reduced pore diameter at time t, the other components being the same as in equation (3). It is interesting to note that in the sintering of glass, at a viscosity range of 10^6 to 10^8 poises, the relaxation times from Fig. 4 would be of the order of 10^{-4} to 10^{-2} sec, which is insignificant compared to the $3.34 \cdot 10^5$ sec employed in the experimental times in sintering of the glass; in this instance, the relaxation constant would impose practically no restraint.

The experimental confirmation of the Frenkel expression at a viscosity range of 10^6 to 10^8 poises with glass can be extended to the melt extrusion or forming of polymers usually made to much lower viscosity

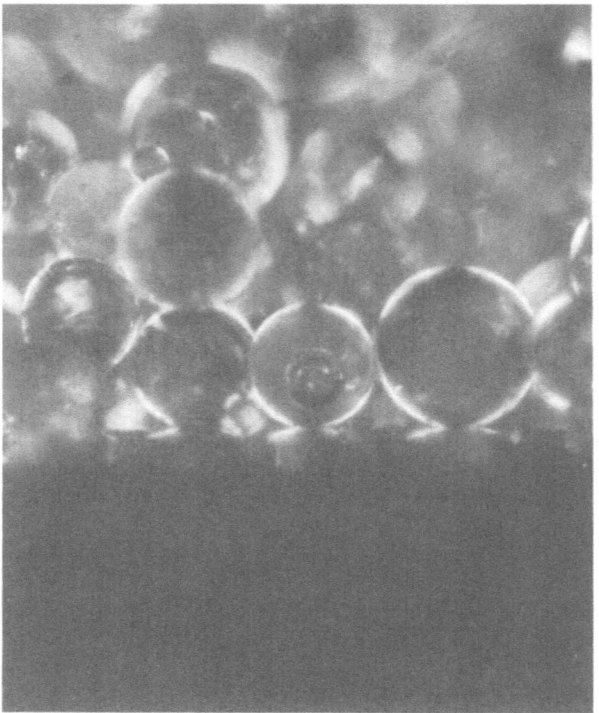

Fig. 6. Sphere of polymethyl methacrylate sintered to a plate after
10 min at 170°C (Kuczynski and Neuville [9]). Magnification 92×

ranges, namely, 10^4 to 10^6 poises. With an average, typical surface
tension of 25 dyn/cm, equation (3) would require about 6 sec for $\frac{1}{16}$-in.-
diameter pellets to complete the initial stage of fusion or where the x/a
quotient would amount to approximately 0.3, beyond which hole closure
by equation (4) would apply and require about 13 sec to close the holes.
These time figures are highly idealized and apply to perfect spheres, but
do not account for irregular particle shapes, as in molding cubes, which
could demand longer times for the first and second stages of fusion.
Additionally, the use of pressure can favor the sintering by reducing the
hole or pore size, but opposing this would be the migration of entrapped
gases by diffusion.

Another common example of coalescence by viscous flow is the consolidation or sintering of latex particles as is practiced in the paint,
paper, and textile technology [8]. Generally, this involves lattices containing 50% of solids dispersed in water and other liquid media. The
average particle size of the latex particles is of the order of 0.1 μ. In

some cases the lattices form films at room temperature, while in others some heat must be applied to effect coalescence. As the water or the liquid medium evaporates, the colloidal particles deposit on the surface, where they fuse into a coherent film. During the evaporation, the viscosity usually starts off from about 10 to 100 poises and gradually increases one or more orders of magnitude. Equation (3) with this initially low viscosity would dictate rapid development of the interparticulate fusion, to be followed by hole closure according to equation (4), where the hole radius, of the order of $r_0 \sim 10^{-5}$ cm, would provide an extremely small time scale to complete the coalescence to continuous film structure. In some cases, fusion of latex mixtures containing dispersed or emulsified solvents can take place by simple solvent action followed by evaporation of the solvent.

VISCOELASTIC COALESCENCE OR SINTERING

In considering polymers that are endowed with viscoelastic characteristics in the viscosity range beyond 10^8 poises, experienced with glass, the mechanism no longer can be equated in simple viscosity terms as indicated by Frenkel's expression [equation (3)]. Kuczynski and Neuville [9] first recognized the inadequacy of equation (3) in extending their glass sintering studies to polymethyl methacrylate used in the form of spherical beads. The technique consisted of sintering the beads to various time intervals, isolating selected fused beads separated carefully to reveal a clear circle, such as shown in Fig. 6. They observed that the data conformed to the general expression

$$\left(\frac{x}{a^{3/5}}\right)^n = K(T)t \tag{5}$$

where K is an experimental constant, T is temperature, t is the sintering time, x is the radius of the interface, and a is the radius of the polymethyl methacrylate sphere. The exponent n decreases from 5 to approximately 0.5 as the sintering temperature increases from 127 to 207° C; neither n nor the particulate a exponent of $\frac{3}{5}$ assume any phenomenological significance. The experimental data are shown in Fig. 7, where none of the indicated exponents n provides any whole integer that would suggest some specific mechanism other than non-Newtonian viscous sintering. Figure 8 shows a rearranged plot of the temperature-dependent exponents of Fig. 6 in relation to the viscoelastic spectrum. It appears that sintering by viscous flow can occur between 168°C [$n = 2.25$ in equation (5)] and 170°C ($n = 1.75$), but this may comprise a balance between two extremes of (a) viscoelastic or restrained flow below this temperature and (b) concomitant degradation of low-viscosity scission products. Figure 7 also purports to show that with increasing temperature the polymer gradually changes from a high-modulus($2 \cdot 10^{10}$ dyn/cm^2) elastic solid at room temperature to a soft viscoelastic material

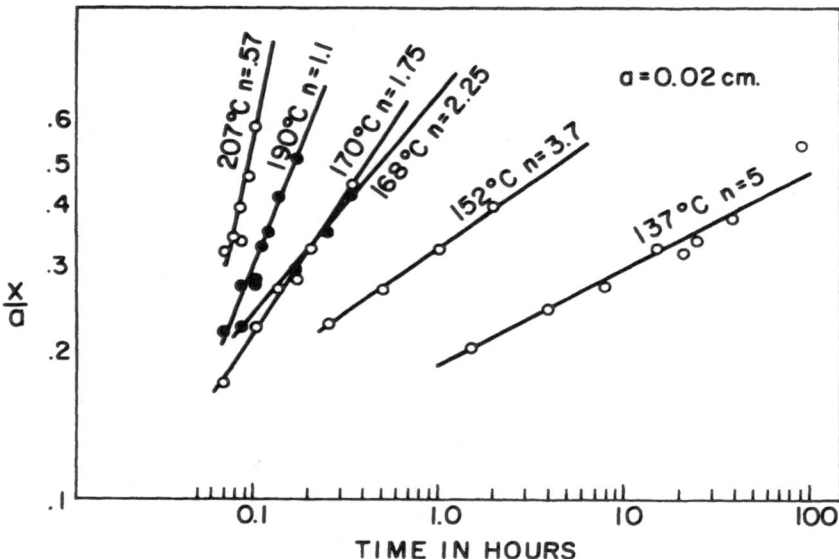

Fig. 7. Time-dependent x/a quotient of polymethyl methacrylate beads sintered at various temperatures (Kuczynski and Neuville [9]).

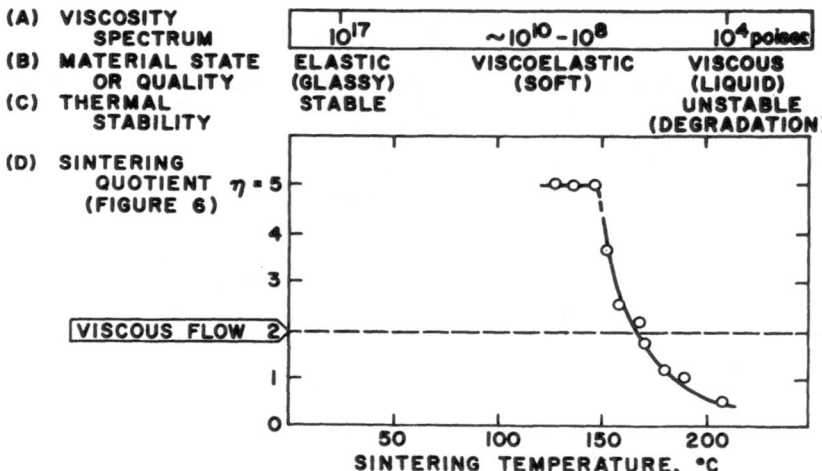

Fig. 8. Sintering quotient for polymethyl methacrylate in relation to viscosity spectrum.

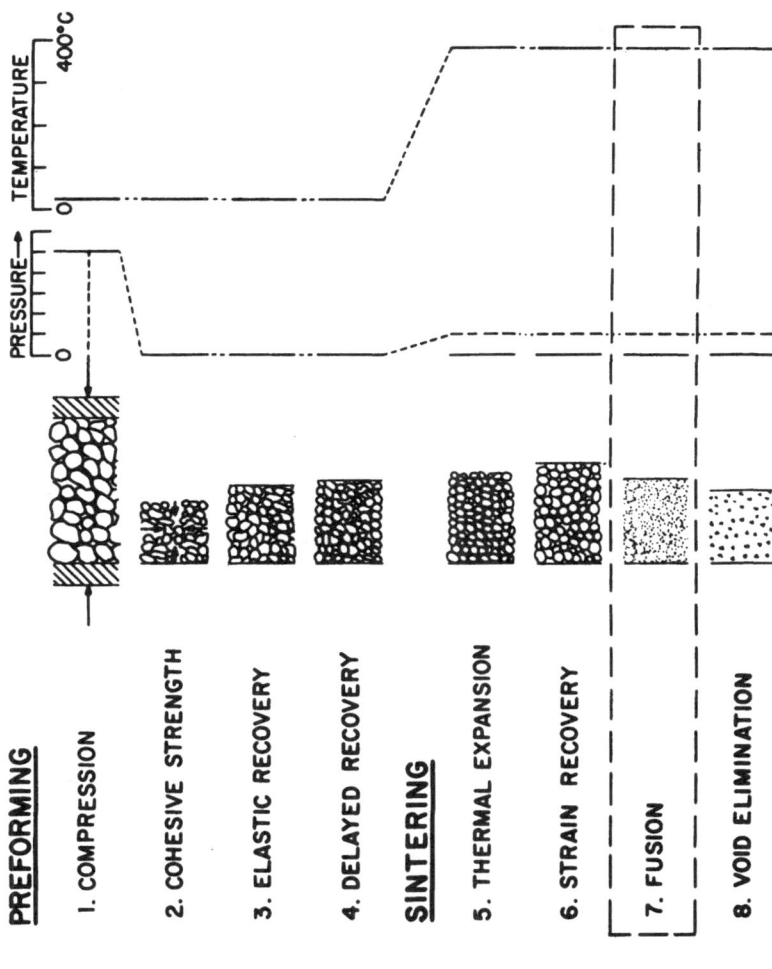

Fig. 9. Schematic diagram of preforming and sintering sequence with polytetrafluoroethylene.

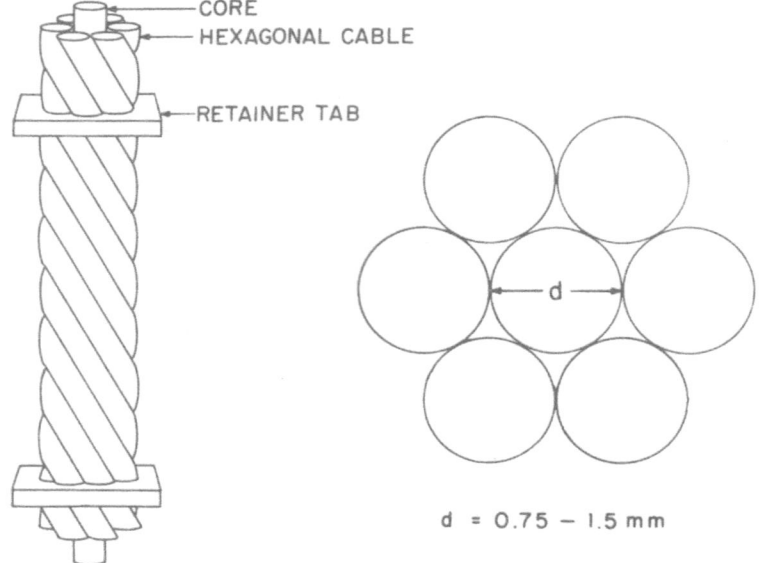

Fig. 10. Schematic diagram of hexagonal arrangement of cable cross section.

Fig. 11. Photomicrograph of microtomed cross section of sintered
polytetrafluoroethylene cable.

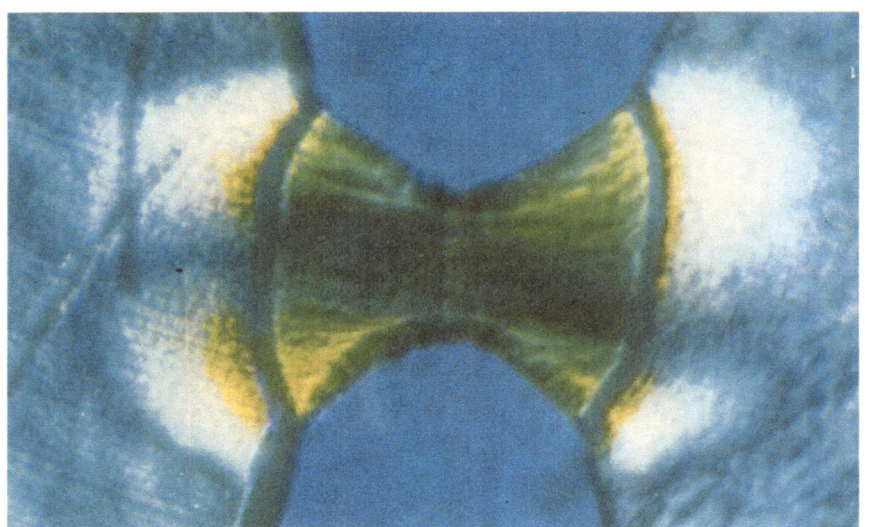

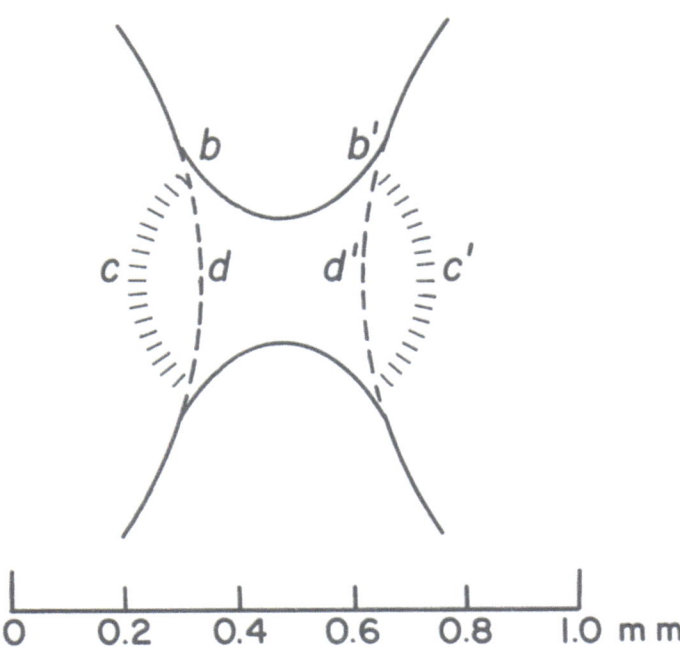

Fig. 12. Photomicrograph of extended sintered junction of polytetrafluoroethylene.

(10^8 to 10^{10} poises) and ultimately to a viscous melt undergoing thermal degradation. In effect, sintering below 168–170°C is controlled or restrained by the elastic character and hence some relaxation constant is needed to account for the actual mechanism. On the other hand, sintering above 168–170°C with the onset of thermal scission of the molecular chains requires a temperature-dependent factor.

Including both of these variants would lead to a most cumbersome modification of equation (3) to be sure. However, for a general understanding of the mechanism it will indeed suffice to apply a suitable retardation constant to Frenkel's expression based on the spherical coalescence. This has been attempted as discussed in the ensuing section on sintering with another viscoelastic polymer, namely, polytetrafluoroethylene, having a viscosity in the range of 10^{10} to 10^{11} poises [5].

SINTERING OF POLYTETRAFLUOROETHYLENE

This polymer is unique for its chemical inertness and heat resistance, but is highly intractable in conventional plastic-molding technology. Its high molecular weight [10] precludes fabrication by usual melt extrusion and other melt-forming methods, except in the case of the corre-

TABLE III

Physical Characteristics of Teflon* 6 Tetrafluoroethylene Resin[†]

Ultimate particle size (average), μ	≈ 0.2
Aggregate particle size, diameter, μ	500 ± 150
Apparent (powder) density, g/liter	475 ± 100
Specific surface (BET), m^2/g	≈ 11
Crystallinity, %	$93-98$
Standard specific gravity	2.2
Apparent melt viscosity, poises	$6 \cdot 10^{10}$
Molecular weight (M_n)	$1-3 \cdot 10^6$

*Teflon is the registered trademark of the Du Pont Company for its fluorocarbon resins.
[†]Type III Resin, American Society for Testing Materials, Specification D-1457-62-T, Table I.

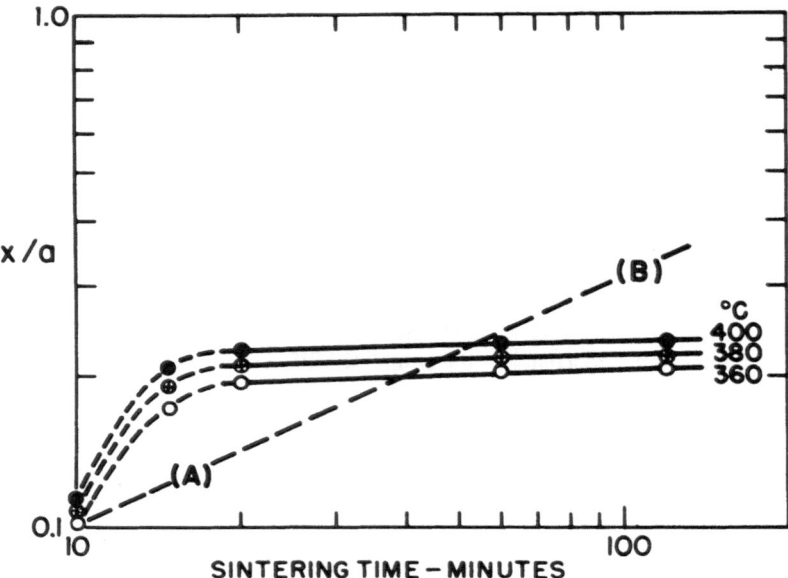

Fig. 13. Time-dependent x/a quotient of sintered polytetrafluoroethylene cross sections.

sponding copolymer with hexafluoropropylene made to the 10^4 to 10^6 poises range [11]. Hence the polymer requires fabrication that more nearly resembles the methods of powder metallurgy technology, namely, (a) preforming the polymer and (b) sintering above its 327°C melt transition temperature. The principal phenomena involved in preforming and sintering comprise first four distinct phases, as shown in Fig. 9: (1) compression, (2) development of cohesive strength, (3) instantaneous or elastic recovery, and (4) delayed recovery; then during sintering additional effects take place that include (5) thermal expansion, (6) strain recovery, (7) interparticulate fusion, and (8) void elimination. Although the effects of preforming and sintering in relation to (7) and (8) have been described by other investigators [12], the basic mechanism involved in actual fusion to a consolidated structure has yet to be defined.

The sintering experiments were carried out on polytetrafluoroethylene using a commercial grade, Teflon* 6 tetrafluoroethylene resin, having the

---

*Teflon is the registered trademark of the DuPont Company for its fluorocarbon resins.

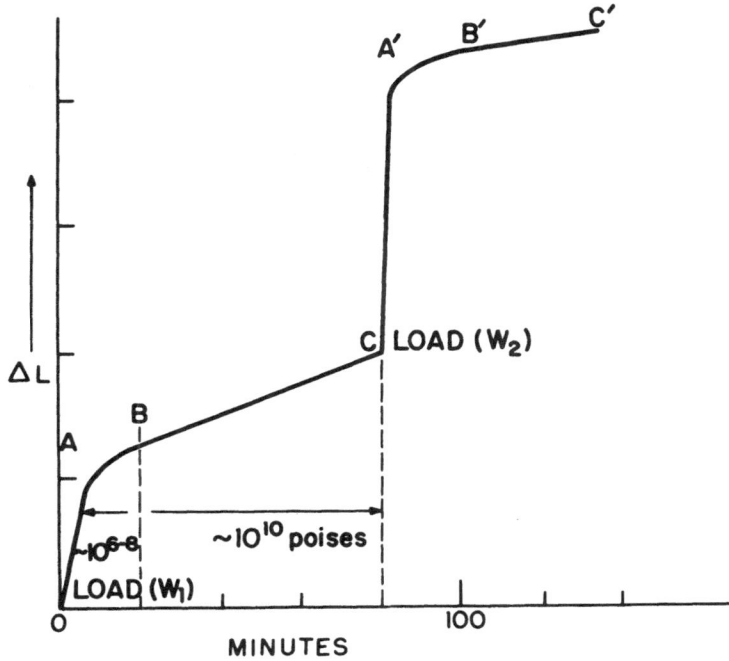

Fig. 14. Typical melt creep curve with polytetrafluoroethylene (Nishioka et al. [5]).

principal physical characteristics shown in Table III. The resin was converted into 30- to 60-mil (0.075–0.15 cm) beading by extruding a resin–naphtha mixture (81/19 parts by weight) according to the method described by Lontz and co-workers[13] and subsequently elaborated upon by others [14,15]. The beading was made up into a cable strand of seven members such as shown in Fig. 10 to form a hexagonal array of contacting cylindrical surfaces as shown in the figure. Following sintering at 360 to 400°C for selected time intervals, the cable strand was cut transversely into microtome sections as shown in Fig. 11 and measured for the x and a dimensions. Evidence for fusion as a result of gross polymer transport, presumably due to segmental diffusion across the boundary interface, can be seen on extending the microtome sections as shown in Fig. 12. It can be noted that the segment $(b–b')$ is drawn approximately 300 μ, thus exposing the interdiffused junction. Under polarized light, birefringent strain patterns can be seen to extend well into the radius (c and c') of the cylindrical boundaries.

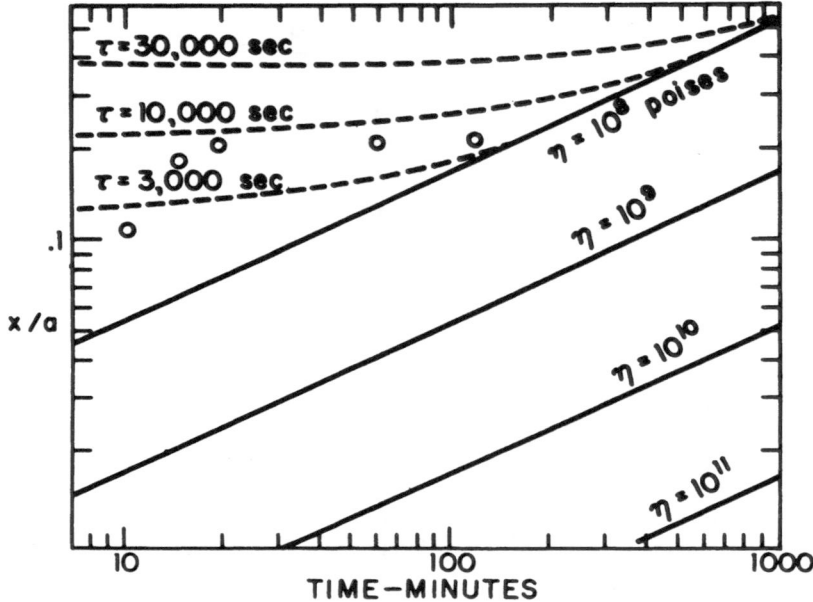

Fig. 15. Comparison of calculated and experimental x/a values: —— From equation (3) —viscous flow mechanism. — — — From equation (2) —viscoelastic flow mechanism. ○ Experimental values at 380°C

Time-Dependent Results

Figure 13 shows graphically the x/a quotient plotted against sintering time at three different temperatures. The absence of a second power slope (AB), shown only as a general line, as required by the Frenkel equation (3) suggests that the sintering of polytetrafluoroethylene involves some mechanism(s) other than that of purely viscous flow. Apparently, a delayed restraint is imposed throughout the 20–120 min time range due to the viscoelastic character of the polymer. An examination of a typical melt creep curve for polytetrafluoroethylene as shown in Fig. 14 [5] provides a plausible support for this concept. Thus, the apparent melt viscosity is discernible from the steady-state slopes BC and $B'C'$ corresponding to a time interval from approximately 20 to 80 min, giving a steady-state flow in the range of 10^{10} poises. However, an appreciable portion of melt creep or flow occurs in less than 20 min between A and B as well as A' and B' with an apparent initial viscosity of the order of 10^6 to 10^8 poises, increasing exponentially to the steady-state level of 10^{10} poises. Hence, this exponential change in viscosity requires an appro-

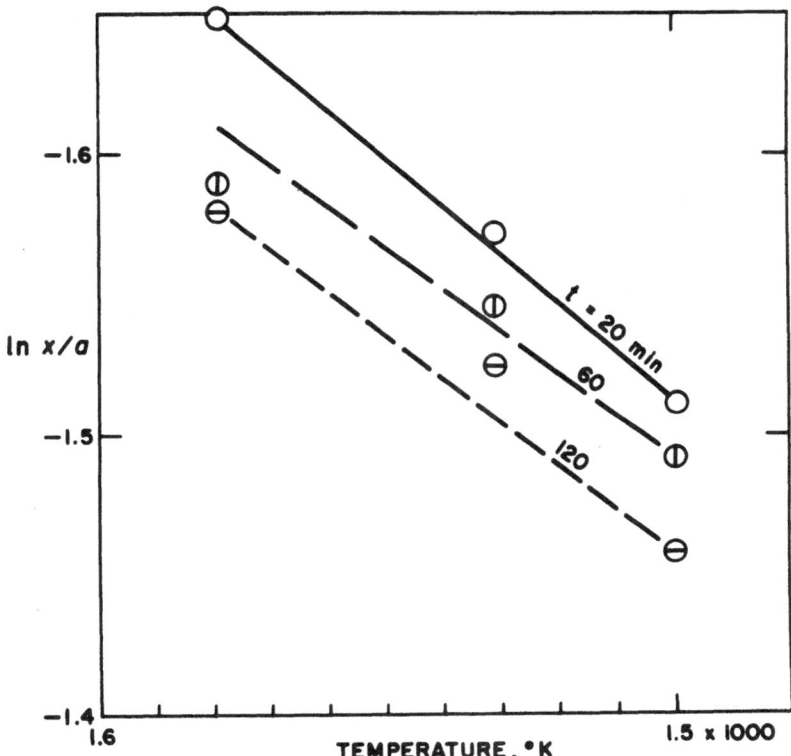

Fig. 16. Temperature-dependent x/a quotient of polytetrafluoroethylene at various sintering times.

priate time constant or relaxation time that can be attributed to either (a) inherent or internal molecular restraints and (b) artificial restraints imposed during the course of the preforming step, or some combination of (a) and (b) acting simultaneously.

The strict application of equation (3) with an arbitrarily assigned surface tension value of 25 dyn/cm^2 and a steady-state viscosity ranging from 10^8 to 10^{11} poises has been found incompatible with the data shown in Fig. 13. It appears that the simple viscosity component in equation (3) must be modified by an appropriate relaxation constant τ in some exponential form as indicated parenthetically in equation (1).

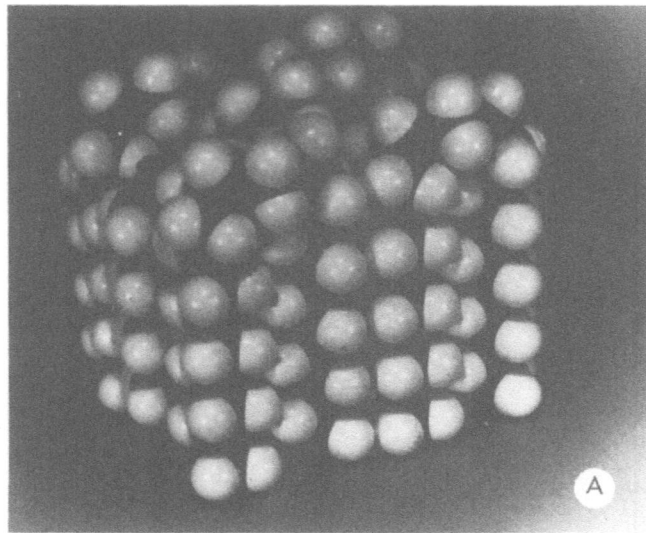

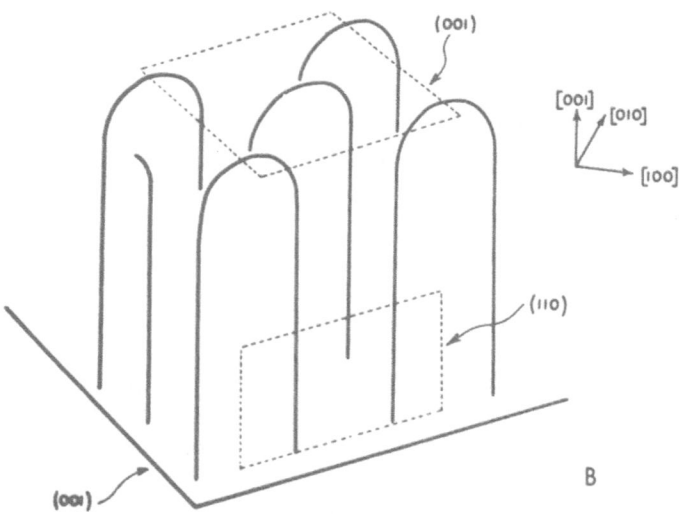

Fig. 17. Folded structure of crystalline organic polymers [from Reneker and Geil, *J. Appl. Phys.* 31:1919 (1960)]. (A) Stuart–Briegleb model showing fold planes resulting in growth of flat lamellar polyethylene crystal. (B) Sketch of (A) showing crystallographic planes and axial directions. (C) Model of lamellar formation

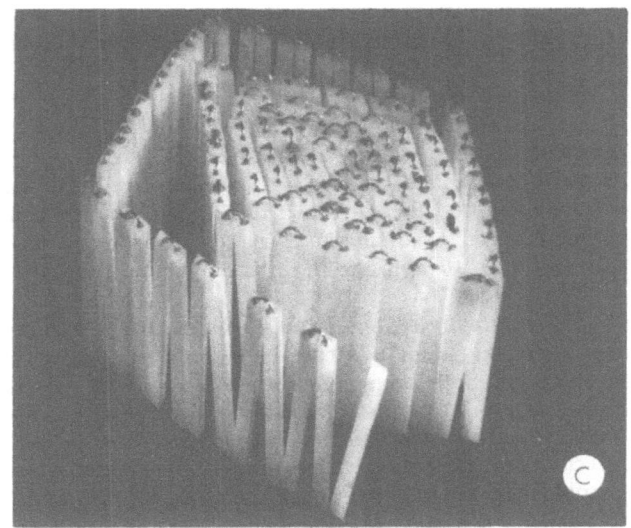

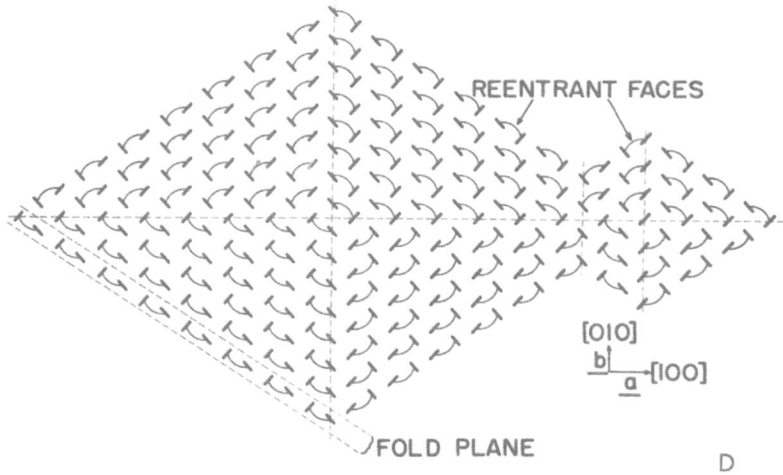

illustrating progressive alignment of folded chains approximately 80 to 120 carbons depending upon crystallization conditions. (D) Sketch of top projection showing arrangement of folds in a crystal. Planar zigzag of molecular backbone. Fold at top of lamella.

Hence it was suggested [16] that for viscoelastic sintering the time-dependent interfacial (x/a) coalescence should be more properly described as

$$\frac{x^2}{a} = \frac{3}{2} \frac{\gamma t}{\eta[1 - \exp(-t/\tau)]} \tag{6}$$

where η represents initial viscosity of order of 10^8 poises and τ an apparent relaxation constant to be assigned from experimental determinations to account for both inherent and stress-induced retardation. Thus from the simple relation of equation (2) with η_3 of order of 10^{11} poises and E_2 of the order of 10^7 dyn/cm^2 [5] τ would amount to $10^{11}/10^7$ or 10^4 sec.

Based on an arbitrarily assigned range of apparent relaxation times, a series of isopleth values for x/a calculated from (a) viscous and (b) viscoelastic sintering have been constructed for Fig. 15 in relation to the experimental x/a values. This is intended to emphasize that the sintering even with assigned viscoelastic restraint constants may start off as a viscous mechanism gradually transforming to one that is viscoelastic. The best graphical fit for the experimental data appears to exist with η equal to 10^8 poises and τ of the order of 10^4 sec.

Temperature-Dependent Results

The data in Fig. 13 indicate the expected increase in the x/a sintered interface with increasing temperature. The Arrhenius temperature dependence, shown graphically in Fig. 16, provides an apparent activation energy of 2–3 kcal, which is considerably lower than the usual range of 24–30 kcal [5] for viscous flow. From Table I it would appear that, instead of segments of large (macro) molecules of the order of 10^2–10^4 A, the steady-state sintering at this low activation energy involves relatively small segments of the order of 50 to 500 A.

CONCLUSION

The sintering or coalescence of polymers can be adequately considered in rheological terms of either viscous or viscoelastic transport or combinations of the two. The classic Frenkel–Kuczynski equation can be applied to the purely viscous mechanism but must be modified in case of viscoelastic sintering with relaxation time constants that can be as high as 10^4 sec, as appears necessary in the case of polytetrafluoroethylene. The two mechanisms may not necessarily be mutually exclusive and could merge into each other either by change in temperature, as appears to be the case with polymethyl methacrylate, or with increasing time of sintering, notably with polytetrafluoroethylene. Thermal degradation of the polymer, as considered in the case of polymethyl methacrylate can have an important effect by changing the viscosity and thus requiring still further temperature- and time-dependent components in the Frenkel–Kuczynski expression.

REFERENCES

1. G. E. Alves and E. W. Burgmann, *Chem. Eng.* 68:181 (1961).
2. H. Mark, *Cellulose and Cellulose Derivatives*, E. Orr, Ed. (Interscience Publishers, Inc., New York, 1943).
3. L. E. Nielsen, *Mechanical Properties of Polymers* (Reinhold Publishing Corp., New York, 1962).
4. A. V. Tobolsky, D. Katz, and A. Eisenberg, *J. Appl. Polymer Sci.* 7:469 (1963).
5. A. Nishioka and M. Watanabe, *J. Polymer Sci.* 28:298 (1957); see also 28:617, 653 (1958) and *J. Appl. Polymer Sci.* 2:114 (1959).
6. J. Frenkel, *J. Phys.* (U.S.S.R.) 9:385 (1945).
7. G. C. Kuczynski and J. Zaplatynskyj, *J. Am. Ceram. Soc.* 39:349 (1956).
8. R. E. Dillon, L. A. Matheson, and E. B. Bradford, *J. Colloid. Sci.* 6:108 (1951).
9. G. C. Kuczynski and B. Neuville, Notre Dame Conference on Sintering and Related Phenomena, June 1950; see also thesis, B. Neuville, "Study of Sintering of Polymethyl Methacrylate," University of Notre Dame, 1958.
10. R. C. Doban, A. C. Knight, J. H. Peterson, and C. A. Sperati, "The Molecular Weight of Polytetrafluoroethylene," Meeting of the American Chemical Society, Atlantic City, September, 1956.
11. Information Bulletin No. X-82a, Teflon 100× Perfluorocarbon Resin, Techniques for Processing by Melt Extrusion, Plastics Department, E. I. du Pont de Nemours & Co., Inc. (1957).
12. A. A. Gorina and V. A. Kargin, *Colloid J.* (U.S.S.R.) (Eng. Trans.) 21:261 (1959); see also *High-Molecular-Weight Compounds*, U.S.S.R., *I*, 1143-7 (1959).
13. J. F. Lontz and W. B. Happoldt, *Ind. Eng. Chem.* 44:1804 (1952).
14. E. E. Lewis and C. M. Winchester, *Ind. Eng. Chem.* 45:1123 (1953).
15. G. R. Snelling and J. F. Lontz, *J. Appl. Polymer Sci.* 3:257 (1960).
16. J. F. Lontz, "Sintering Studies on Polytetrafluoroethylene," Fourth Delaware Valley Regional Meeting, Philadelphia, Pennsylvania, American Chemical Society, January, 1962.

Plastic Deformation

Plastic Deformation in Metals

G. S. Ansell

Department of Materials Engineering, Rensselaer Polytechnic Institute, Troy, New York

INTRODUCTION

One aspect of materials transport of prime interest to materials scientists and engineers is the response of metals to an applied stress. Under the action of an applied stress, metals deform and thus change shape. The nature of this deformation is shown schematically in Figs. 1, 2, 3, and 4.

The behavior illustrated in Figs. 1 and 2 would be typical of a situation where a constant strain rate is imposed and the resultant stress measured as a function of strain.

Figure 1 shows the relationship between applied tensile stress and tensile strain or change in length per unit length, observed generally for nonferrous metals and alloys. In this figure, it can be seen that as the strain increases, the stress increases rapidly, and is approximately proportional to the strain. At some strain level, however, the strain increases with very little increase in stress. The following terminology has been utilized to describe this behavior: The slope of the initial, approximately linear, portion of the stress–strain curve is called Young's modulus, E. That stress above which stress is no longer proportional to the strain is called the proportional limit. The stress at which the total strain is some given amount, usually 0.02 or 0.2% greater than would be observed if the stress remained proportional to the strain, is called the offset yield stress. The increase of stress with increasing strain beyond the yield stress is said to be due to work or strain hardening.

Figure 2 shows the stress–strain behavior observed for some ferrous alloys. As contrasted with the behavior shown in Fig. 1, it is seen that the stress and strain remain approximately proportional until a stress level is reached where the strain increases rapidly accompanied by a decrease in stress. This stress is called the upper yield stress. The lower stress value observed after yielding occurs is the lower yield stress.

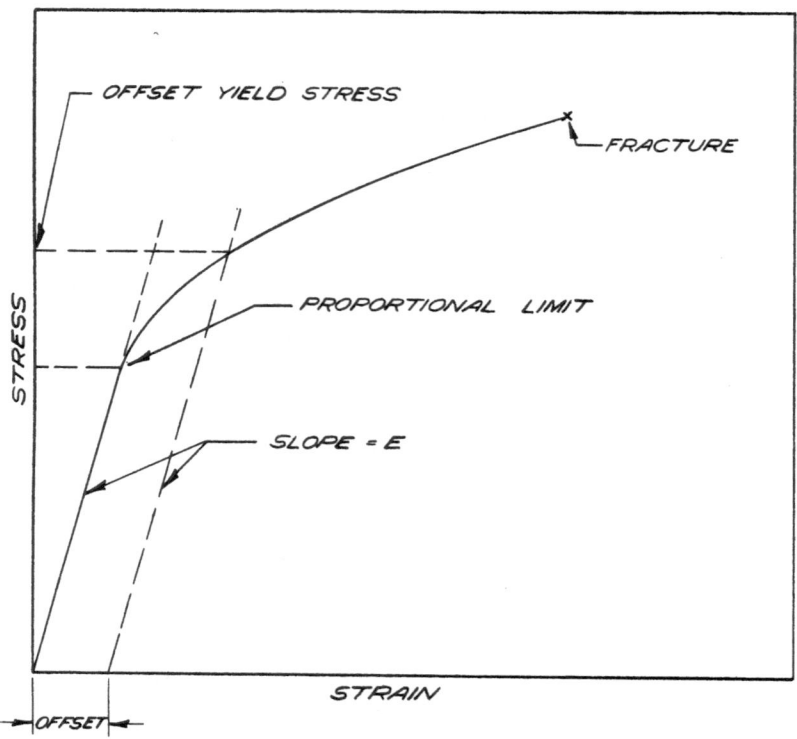

Fig. 1. Schematic plot of stress–strain behavior generally observed for metals.

In some applications, however, a constant value of stress is applied to a metal and the resultant strain is measured as a function of time. This is called creep behavior. Figure 3 schematically shows this behavior for metals at temperatures less than one-half of their absolute melting temperature. It can be seen in this figure that the slope of this curve, or strain rate, $\dot{\epsilon}$, constantly decreases with time.

Figure 4 schematically illustrates this behavior for metals at temperatures greater than one-half of their absolute melting temperature. In this figure it is seen that the strain rate does not constantly decrease, but rather exhibits three different responses. Initially the strain rate decreases with time. This region of the creep curve is termed transient creep. The creep rate then remains essentially constant. This portion of the creep curve is called steady-state creep behavior. In the third or tertiary creep range, the creep rate increases with time until fracture occurs.

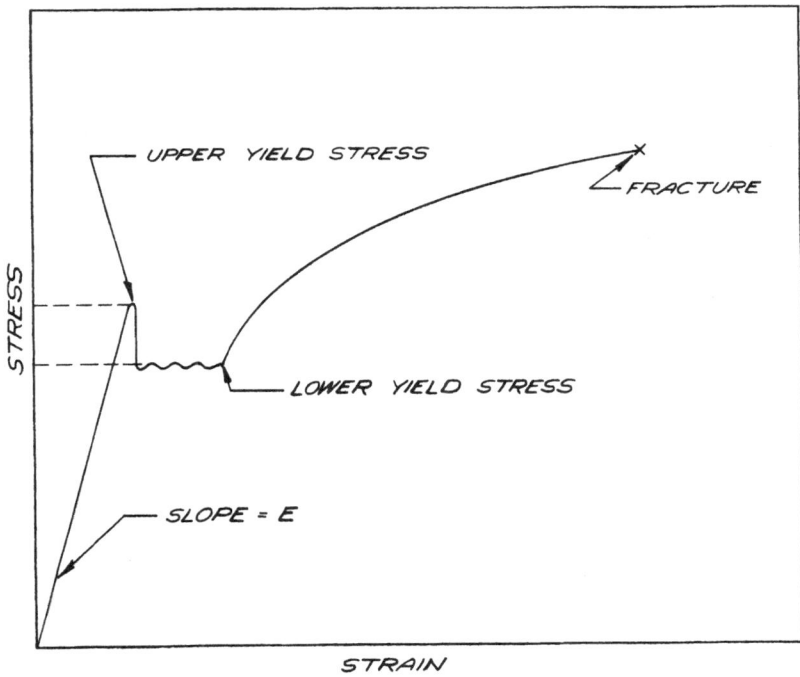

Fig. 2. Schematic plot of stress–strain behavior generally observed for some ferrous alloys.

The deformation shown in these four figures consists of three distinct parts: elastic, anelastic, and plastic strain. Elastic deformation occurs instantaneously under the application of an applied stress and remains only while the stress is maintained. When the stress is removed, the elastic strain disappears or recovers instantaneously. This elastic deformation is reflective of the resistance required to change the interatomic spacing in the metal. Its magnitude is approximately proportional to the applied stress. In Figs. 1 and 2, the strain associated with the initial, approximately linear, portions of the stress–strain curves is largely elastic strain. In Figs. 3 and 4, the instantaneous strain shown at loading at time equal to zero is elastic strain. Anelastic deformation is a time-dependent process which starts upon the application of stress. When the stress is removed, the anelastic strain recovers completely over some period of time. The magnitude of anelastic strain in metals generally is small. As contrasted with these two recoverable parts of deformation, the plastic strain remains even after the applied stress is removed. This

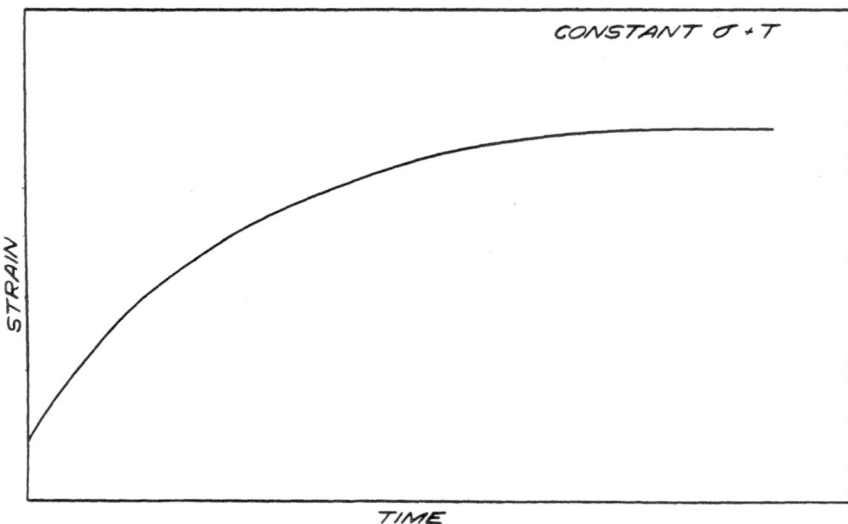

Fig. 3. Schematic plot of low-temperature creep behavior.

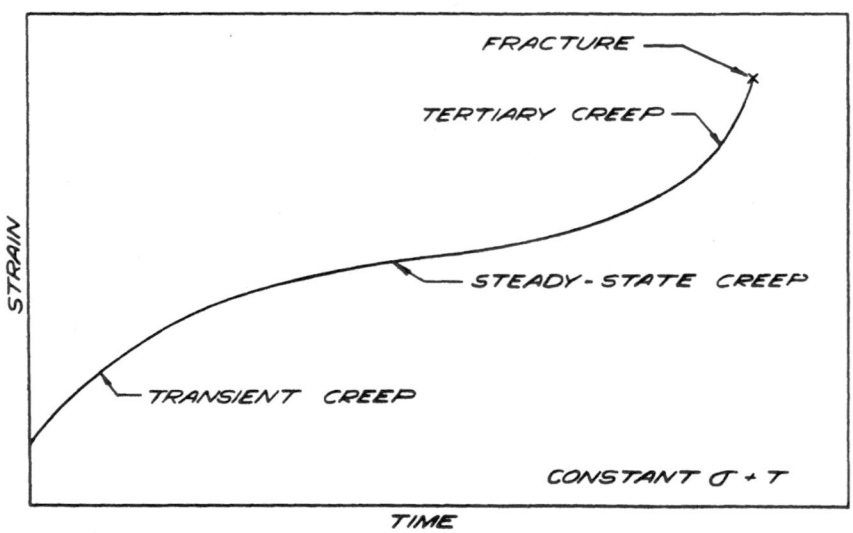

Fig. 4. Schematic plot of high-temperature creep behavior.

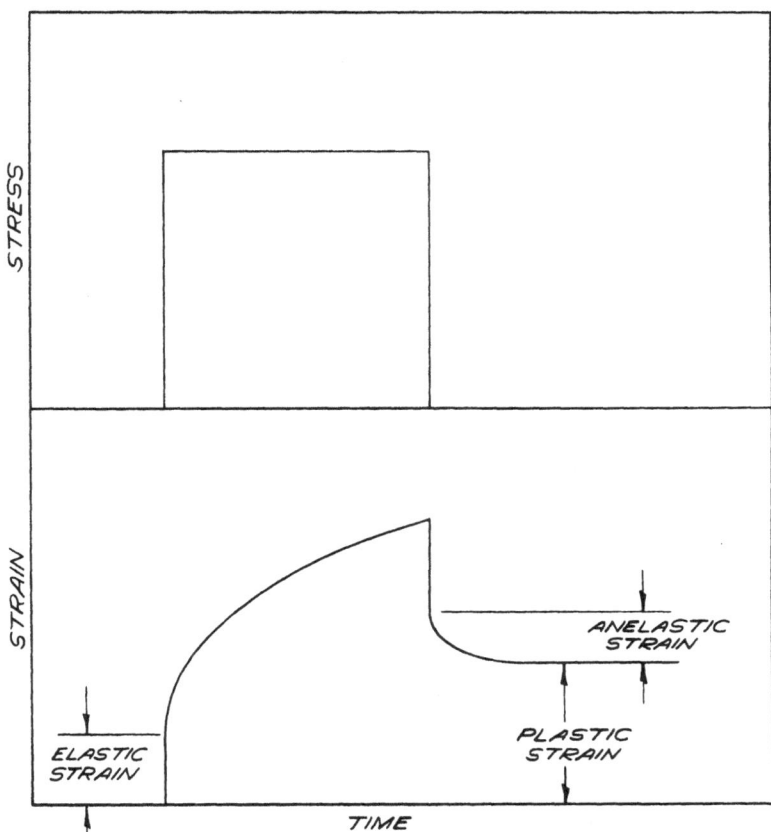

Fig. 5. Schematic plot of applied stress and resultant strain as a function of time.

permanent, time-dependent strain is the predominant type of deformation observed when metals are subjected to stresses higher than the yield stress in tensile deformation and for the time-dependent creep deformation. The distinction between these types of deformation can, perhaps, be seen more clearly in Fig. 5. In this figure, applied stress and resultant strain are shown as a function of time. On this schematic plot, each type of strain is shown.

For many engineering applications, the utilization of metals for structural purposes requires the maintenance of some degree of dimensional stability. At low temperatures, relatively little strain occurs at stresses below the yield stress. At higher temperatures, the transient and steady-state creep rates dictate service life. Even when dimensional stability is not critical, the yield stress serves to delineate the mechanical strength of metals. On this basis the parameters indicated on the first four figures have been widely utilized for determining the design application of metals and alloys. Underlying these parameters is, however, the resistance to the initiation and propagation of plastic deformation. For this reason, the nature of plastic deformation in metals will be emphasized in the remainder of this paper.

APPROACH

The rationale for the study of plastic deformation in metals arises principally from the need for predictability concerning the mechanical behavior of metals in specific structural applications. The approaches utilized to satisfy this need have taken four basically different routes, which are all in use at the present time, each with its distinct advantages and disadvantages. They are (1) direct test (2) parametrical specification, (3) rheological description, and (4) applied dislocation theory.

Direct Test

There is perhaps no more accurate method of determining the performance of a material than to try to utilize it in its specific application. This is, of course, the ultimate test. No matter what previous work is performed, eventually this method is utilized. As an initial design indication it is, however, expensive, sometimes dangerous, and time-consuming, and must be performed separately for every application. There is no need to dwell longer on this most valid and self-explanatory of all routes.

Parametrical Specification

Rather than test materials in a specific application, it is more efficient to first analyze the specific design environment, e.g., state of stress and temperature, and the level of mechanical response required, and then to base the materials selection upon parametrical responses received from materials as determined in standard test procedures which more or less stimulate the actual service environment. These parametrical responses may be interpreted in terms of correlation made with service performance in specific applications. Hence the widely disseminated values for such quantities as modulus, yield strength, and steady-state creep rate for engineering materials. The degree of success of this method is apparent from its wide application, and also from the extensive safety factors concomitant with its use.

Rheological Description

Inherent in each of the above routes toward handling the mechanical response of metals is the utilization of extensive testing programs, generally an expensive proposition. In order to avoid such expensive procedures, it has been proposed that the mechanical behavior of all metals be described mathematically in equations of similar form, where the constants in these equations may be more easily determined for each metal and alloy; e.g., the stress may be written as the product of Young's modulus and elastic strain for tensile deformation under low stresses. The optimal result from such a procedure would be to write equations analogous to thermodynamic equations of state, where, if all but one of the variables—stress, strain, strain rate, and temperature—are specified for a given material, the remaining variable would be uniquely determined. The rationale leading to this approach lies in the observation of the following type of data:

a. When temperature and strain rate are held constant, e.g., from a tensile test, at stresses above the yield stress, it is often observed that

$$\log \sigma = (\log \epsilon)^{m}$$

where σ is stress, ϵ is strain, and m is called the strain-hardening exponent and varies between 0.01 and 0.8 for most metals.

b. When temperature and strain are held constant, e.g., from a series of tensile tests, or from a tensile test in which the strain rate is changed, it is often observed that

$$\log \sigma = (\log \dot{\epsilon})^{n}$$

where $\dot{\epsilon}$ is strain rate and n is called the strain-rate exponent and varies between 0.001 and 0.2 for most metals.

c. When stress and strain are held constant, e.g., when temperature is abruptly changed in a creep test, it is often observed that

$$\dot{\epsilon} = A e^{(-Q/kT)}$$

where A is a constant and Q has the same dimensions as an activation energy.

From these relationships, if they are unique for a particular metal or alloy, one should be able to write an equation of state of the type

$$\sigma = K^{T}(\dot{\epsilon}_{1}/\dot{\epsilon}_{2})^{BT} \epsilon^{f(T)}$$

in which each of the constants, K and B, can be quickly determined for a particular metal or alloy from a series of relatively simple tests. Thus, more complex behavior might be simply calculated.

A graphical example showing the advantage of this type of prediction method is readily seen in the example of predicting creep behavior, ordinarily expensive to determine, from relatively simple tensile behavior as shown in Figs. 6, 7, 8, and 9.

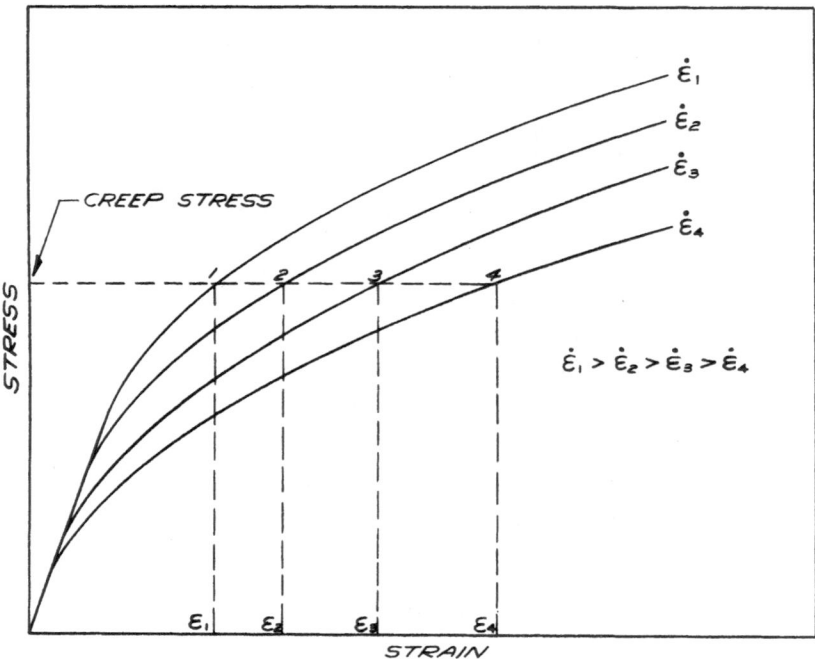

Fig. 6. Schematic plots of stress–strain behavior determined at various strain rates.

As shown in Fig. 6, a series of tensile curves is determined for various strain rates at the same temperature as desired for the predicted creep behavior. From these data, the strain at each strain rate is measured for the stress value desired for the predicted creep behavior. The strain rate is then plotted as a function of these measured strains, as shown in Fig. 7. It is relatively safe to extrapolate this curve beyond the plotted data points. This extrapolated curve is then replotted in Fig. 8 as inverse strain rate as a function of strain. The integral of this plotted function, or the area under the curve, is time, since

$$t = \int_0^\epsilon \frac{dt}{d\epsilon} d\epsilon$$

Therefore, from this curve the time to reach a given strain can be determined, and if an equation of state is applicable, the creep curve shown in Fig. 9 may be derived.

Unfortunately, however, extensive research has shown that such an

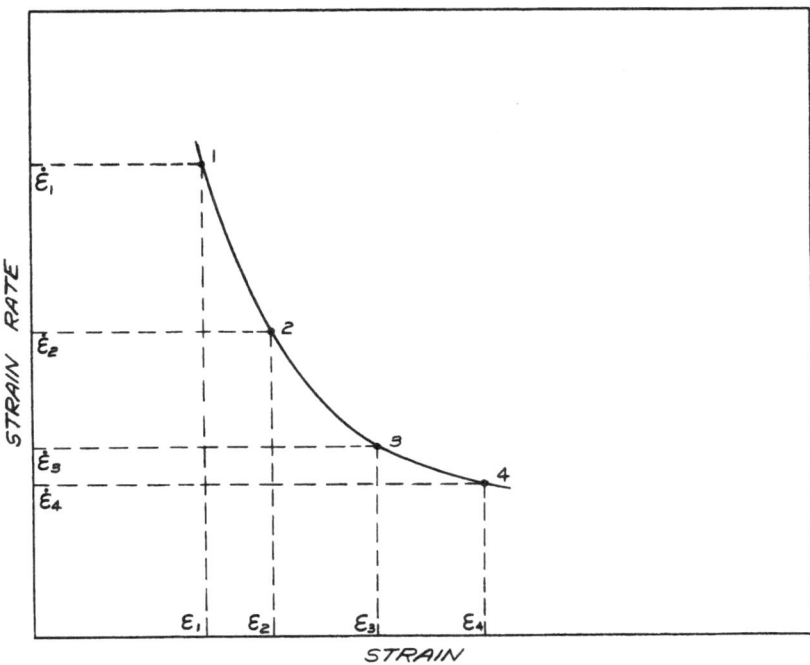

Fig. 7. Data from Fig. 6, plotting strain rate vs. strain at constant stress.

equation of state applies to mechanical response only if no metallurgical or substructural changes occur in the metal. Thus, such a treatment is not generally valid, and can apply only over restrictive variable ranges. For this reason the concept of such a mechanical equation of state has fallen into disrepute.

Without bearing the onus of the terminology "mechanical equation of state," other rheological descriptions have been empirically determined to predict mechanical behavior of metals. These carry various labels, e.g., stress rupture parameters. It should be sufficient at this point to recall that "a rose by any other name...."

Applied Dislocation Theory

This last route toward handling the mechanical response of metals is based upon the quantitative evaluation of the relationship of metallurgical and lattice defect structures with the fundamental aspects of the plastic deformation process.

In such a treatment, one must first consider the nature of plastic deformation in metals. Taylor [1], Orowan [2], and Polanyi [3], during

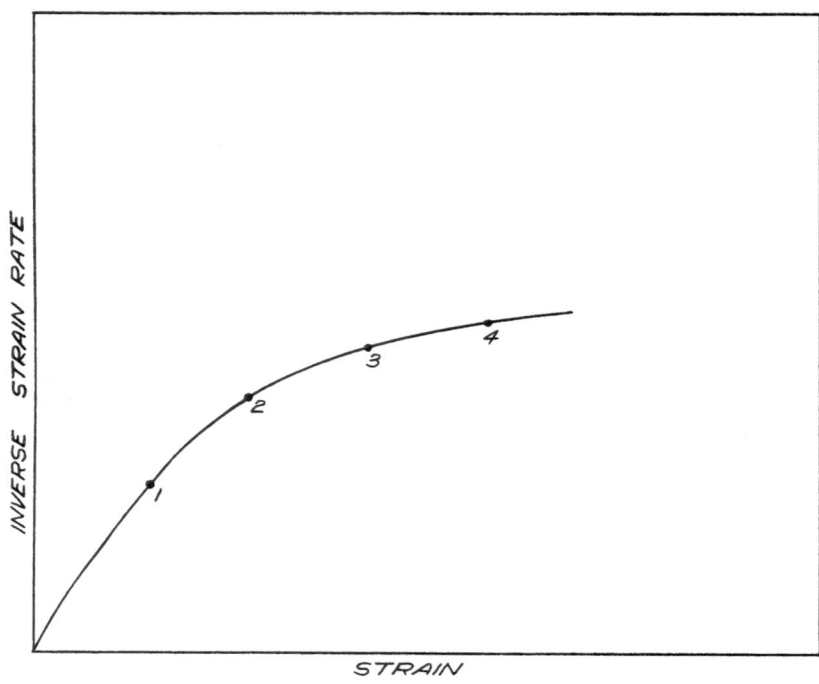

Fig. 8. Inverse strain rate plotted vs. strain using data shown in Fig. 7.

the early 1930's, realized that the fundamental nature of plastic deformation had to be compatible both with gross observations of deformation
and with the nature of the crystalline structure and interatomic binding
forces present in metals. Their work led to the concept of crystalline
plasticity, termed dislocation theory.

Plastic deformation in crystalline solids can take place only by slip,
twinning, or diffusion. In metals, slip or deformation due to the movement of dislocations is generally the least restrictive and most important
plastic deformation process. This process is relatively inhomogeneous.
Slip is the translation of the crystalline lattice across a crystallographic
plane in a crystallographic direction. In metals, the slip planes are usually the family of planes with closest atomic packing, while the slip
directions are the directions of densest packing within these plains; e.g.,
in fcc lattices (111) planes and <110> directions. From energy considerations it was quickly seen that this translation could not occur simultaneously across an entire crystallographic plane, but rather starts at
some point in the plane and then spreads out over the plane. In a plane

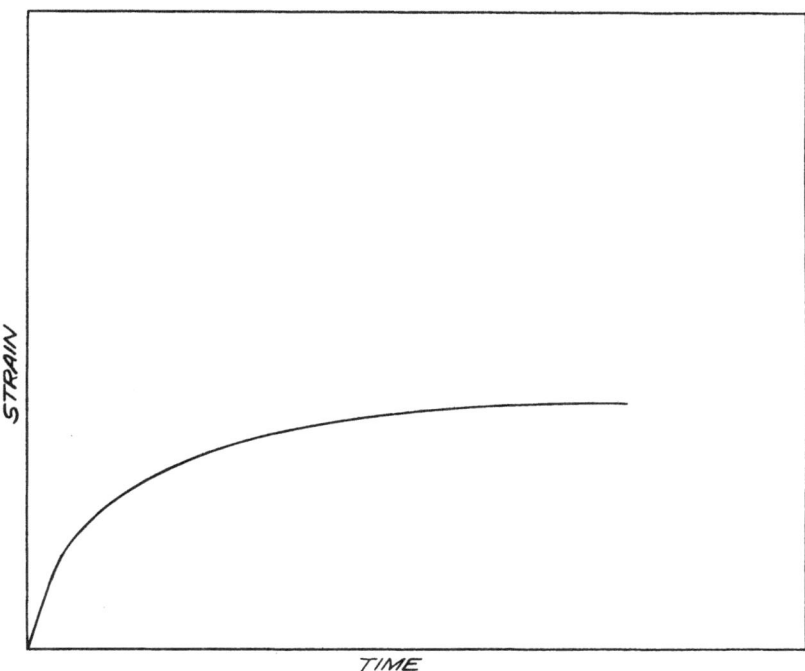

Fig. 9. Creep curve derived from Fig. 8.

where slip is occurring, there is therefore a portion of the plane where the slip translation has occurred surrounded by an as yet unslipped area. Both the slipped and unslipped portions of the slip plane are part of normal crystalline lattices. The boundary separating these areas must therefore be one of accommodation and is called a dislocation. Figure 10 is a schematic view of a plane in which slip is taking place.

The portion marked A has slipped, while the remainder of the plane is unslipped. The boundary separating these regions is a dislocation. Although the shape of the dislocation line is dictated simply by the shape of the slipped area, the dislocation line may be characterized by the amount of translation it accommodates, since the direction and amount of slip within one area of slip must be an invariant. This translation or slip vector is called the Burgers vector of the dislocation. The Burgers vector completely characterizes the dislocation. Since the structure of a dislocation is necessarily one of accommodation, the dislocation is a crystalline defect whose structure depends upon the crystalline lattice, the Burgers vector, and the orientation of the dislocation relative to the

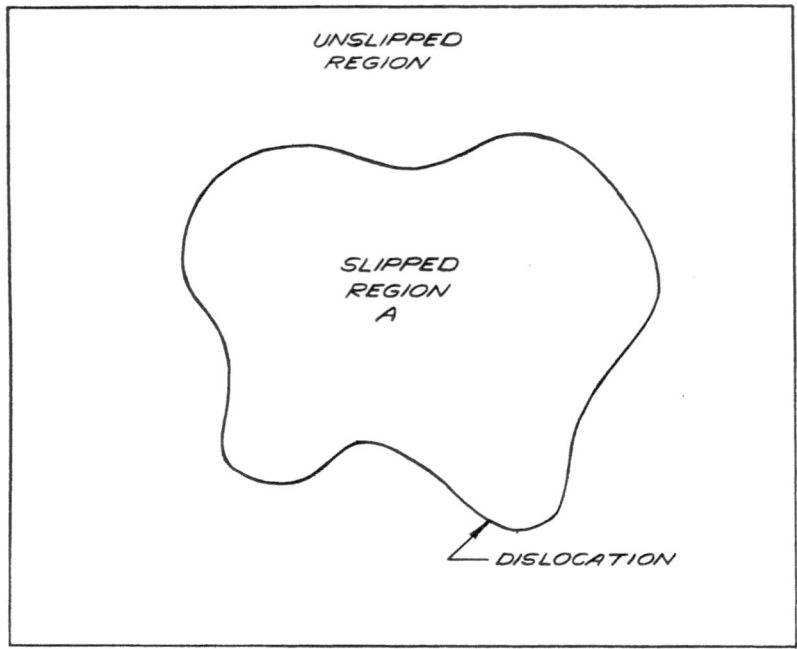

Fig. 10. Schematic view, looking down on slip plane.

direction of the Burgers vector. In order to preserve the crystalline nature of the metallic lattice, the magnitude of the Burgers vector must be an integral multiple of the atomic spacing in the slip direction. In metals, the energy of the dislocation arises primarily as a result of the elastic strain energy associated with the structural accommodation present in this boundary. One can show that this energy is proportional to the square of the Burgers vector. For this reason, the magnitude of a unit slip translation is one atomic spacing. One can, therefore, describe the slip process and hence plastic deformation in metals simply by detailing the motion of the dislocations. The strain in a crystal can be written as

$$\epsilon = \rho b l$$

where ρ is the length of dislocation line per cubic centimeter, b is the Burgers vector of the dislocations, and l is the average distance moved by each of these dislocations, while the strain rate can be written as

$$\dot{\epsilon} = MNAb$$

where M is the density of dislocation sources, A is the area swept out by each dislocation, and N is the rate of generation of dislocations from each source.

From evaluation of the strain field associated with a dislocation, it has been possible to determine the interaction of external stress and internal structure with the dislocation structure in a metal. Thus it is possible to calculate the gross mechanical response of a metal if the metallurgical and defect structure can be delineated. For want of better terminology, this type of fundamental treatment may be called "model calculations." It should be pointed out here, however, that unlike the empirical treatment used in the rheological method, the model calculations inherently consider structural changes.

The nature of such model calculations can be seen best by treating the mechanical behavior of a metallurgical structure by this means. One example of this technique is the calculation of the yield strength of dispersion-strengthened alloys proposed by Ansell and Lenel [4, 5].

Their treatment is as follows: Plastic deformation in crystalline solids is due to the movement of dislocations. Crystals yield when large numbers of dislocations can move appreciable distances through the lattice. This process requires that under the action of an applied stress, two consecutive processes must occur: first, dislocations are generated at some source; and second, these dislocations move appreciable distances through the crystal. Whichever of these requires the higher stress—dislocation generation or dislocation mobility—controls the yielding behavior of the solid. In single-phase materials, either of these two processes may be controlling. In pure annealed metals, dislocation generation generally controls the yielding behavior. In solid-solution-strengthened metals and in some nonmetallic crystals, dislocation mobility may be more restrictive.

In dispersion-strengthened alloys, however, the stress necessary to move dislocations appreciable distances along a lattice plane is higher than the stress required to generate dislocations from a source. In this case, the interaction of the dislocations with the dispersed second-phase particles controls dislocation mobility and hence the yielding behavior.

Model: Dislocations are formed at dislocation sources under the action of an applied stress. The nature of the sources is not critical in considering the model. As the dislocations expand from sources, they are either blocked from further motion by the second-phase particles, or they continue to move by bowing about the dispersed particles, leaving residual dislocation loops surrounding each particle. In either event, if the dislocation is completely blocked by the particles, or if the dislocations bow past the particles, yielding does not result.

Even at one half or more of the absolute melting temperature of the matrix metal, fracture or shear of the second-phase particles should be necessary for appreciable yielding to occur unless recovery takes place. Recovery can occur either by climb of piled-up dislocations at a rate exceeding the applied strain rate, or by cross-slip of piled-up dislocations out of the slip plane, if the geometry of the dispersed-phase particles permits. The possibility of recovery is not considered in the following calculations.

Calculations Based Upon the Model: On the basis of the preceding model, the yield strength of a dispersion-strengthened alloy is now evaluated [4, 5].

The method of calculation is straightforward. First the shear stress τ on the dispersed-phase particle, due either to the lead dislocation from the dislocation source or to the inner residual dislocation loop if the lead dislocation has bowed past the dispersed particle, is calculated as a function of the externally applied stress on the crystal σ. Then the shear stress F required to either shear or fracture the dispersed particle, is evaluated. From these calculations the yield stress of the dispersion-strengthened alloy or the externally applied stress required to shear or deform the dispersed particles, is determined.

The shear stress τ on the dispersed-phase particle is

$$\tau = n\sigma \tag{1a}$$

where n is the number of dislocations piled up against or looped around the dispersed particle, if the radius of curvature of the dislocation nearest the particle is greater than $\mu b/\sigma$, where μ is a shear modulus of the matrix crystal ($\mu = [\tfrac{1}{2}C_{44}(C_{11} - C_{12})]^{\frac{1}{2}}$ for cubic crystals, C_{ij} being the usual elastic constants).

If the radius of curvature of the dislocation nearest the particle is less than $\mu b/\sigma$, the shear stress τ on the dispersed-phase particle is

$$\tau = n\mu b/r \tag{1b}$$

where r is the radius of curvature of the dislocation nearest the dispersed particle.

The number of dislocations, n, in equations (1a) and (1b) acting on the particle depends upon the spacing between particles, according to the relation

$$n = 2\lambda\sigma/\mu b \tag{2}$$

where λ is the spacing between dispersed-phase particles.

When equations (1a) and (2) are combined, the shear stress on relatively large dispersed-phase particles, i.e., in the case of spherical particles, particles whose diameter d is greater than $2\mu b/\sigma$, is

$$\tau = 2\lambda\sigma^2/\mu b \tag{3a}$$

If equations (1b) and (2) are combined, the shear stress on relative fine dispersed-phase particles, i.e., in the case of spherical particles, particles whose diameter d is less than $2\mu b/\sigma$, is

$$\tau = 2\lambda\sigma/r \tag{3b}$$

The shear stress F that will either shear or fracture the dispersed-phase particle is, in general, proportional to a shear modulus μ^* of the particle. Therefore

$$F = \mu^*/C \qquad (4)$$

where C is a constant of proportionality, which can be shown theoretically to be in the neighborhood of 30. For some types of dispersed particles, particularly in the case of the metastable phases or zones which appear as the result of the heat treatment of some precipitation-hardened alloys, other methods [6] of evaluating the shear strength F of the particles have been utilized. If a system is considered where such an alternative estimate of the shear strength appears warranted, its substitution for the shear strength predicted by equation (4) would be justified.

If equations (3a) and (4) are combined, the yield strength, σ_{YS} of a dispersion-strengthened alloy containing relatively coarse dispersed-phase particles is

$$\sigma_{YS} = (\mu b \mu^*/2\lambda C)^{\frac{1}{2}} \qquad (5a)$$

If equations (3b) and (4) are combined, the yield strength, σ_{YS} of a dispersion-strengthened alloy containing fine dispersed-phase particles is

$$\sigma_{YS} = \mu^* r/2\lambda C \qquad (5b)$$

Since it can be shown that in the case of an alloy containing spherical particles that

$$(\lambda + d)/d = 0.82/f^{1/3} \qquad (6)$$

where f is the volume fraction of second-phase particles, equation (5b) may be rewritten in the form

$$\sigma_{YS} = \frac{\mu^*}{4C} (f^{1/3}) \left(\frac{1}{0.82 - f^{1/3}} \right) \qquad (7)$$

When the dispersed-phase particles are not spherical, it becomes necessary to judiciously evaluate the particle size which delineates the boundary between the use of the computations for either the relatively coarse or fine particles. The two computational treatments are of course compatible with the model, each being the limiting case of the other.

These calculations provide both the quantitative prediction of the yield strength of existent dispersion-strengthened alloys and the structural detailing necessary to provide a basis for the development of alloys for specific strength application.

This structure–property concept applied to model calculations has been successfully utilized in delineating several other aspects of plastic deformation [7, 8]. This concept is, however, still in its infancy and it is to be expected that more extensive application of the model method will be undertaken in the future.

From this brief discussion of plastic deformation in metals, it can be seen that the materials scientist has reached the point where the gross mechanical behavior of metals can now begin to relate to the fundamental nature of the metallic state and aggregate structure.

REFERENCES

1. G. I. Taylor, *Proc. Roy. Soc.* A145:362 (1934).
2. E. Orowan, *Z. Phys.* 89:605, 614, 634 (1934).
3. M. Polanyi, *Z. Phys.* 89:660 (1934).
4. G. S. Ansell and F. V. Lenel, *Acta Met.* 8:612 (1960).
5. G. S. Ansell, *Acta Met.* 9:518 (1961).
6. A. Kelly and M. E. Fine, *Acta Met.* 5:365 (1957).
7. G. S. Ansell and J. Weertman, *Trans. AIME* 215:838 (1959).
8. J. Weertman, *J. Appl. Phys.* 26:1213 (1955).

Dislocations in Ceramic and Metal Crystals

J. J. Gilman*

Brown University, Providence, Rhode Island

INTRODUCTION

Although crystal dislocations were invented to explain the plastic behavior of metals and most of the early theory was developed with metal crystals in mind, it seems clear that the final stages of research and development of knowledge about dislocations will be concerned with their behavior in ceramic crystals.

In the case of metals, the plastic behavior is related to the elastic properties of dislocations because details of the crystal structure do not strongly affect the atomic cohesion. The elastic properties of a dislocation describe the region surrounding the center, but not including the central few atoms. Since metallic binding forces are not sensitive to the exact atomic arrangement at the dislocation's center, the energy density at the center is not very different from what it is in the immediately surrounding elastic region, and the behavior is not sensitive to changes at the center.

However, the situation is very different in ceramic crystals because they have very specific structures that are based on definite coordination numbers and distinct bond angles. Also, they are polyatomic rather than monatomic, and this severely restricts the feasible geometric arrangements at the cores of dislocations. Therefore, the energy density in the core region may be quite different from that in the elastic region. The resulting differences of dislocation behavior in ceramics compared with metals cause some features of their mechanical behavior to be quite different. In particular, ceramic crystals tend to be much more brittle than metallic ones.

Before I discuss some comparisons between metallic and ceramic crystals, I shall briefly review the dislocation mechanism of plastic flow in order to define the pertinent parameters. Figure 1, which may already be familiar, shows that a plastic change of shape starts at a certain

---

*Present address: University of Illinois, Urbana, Illinois.

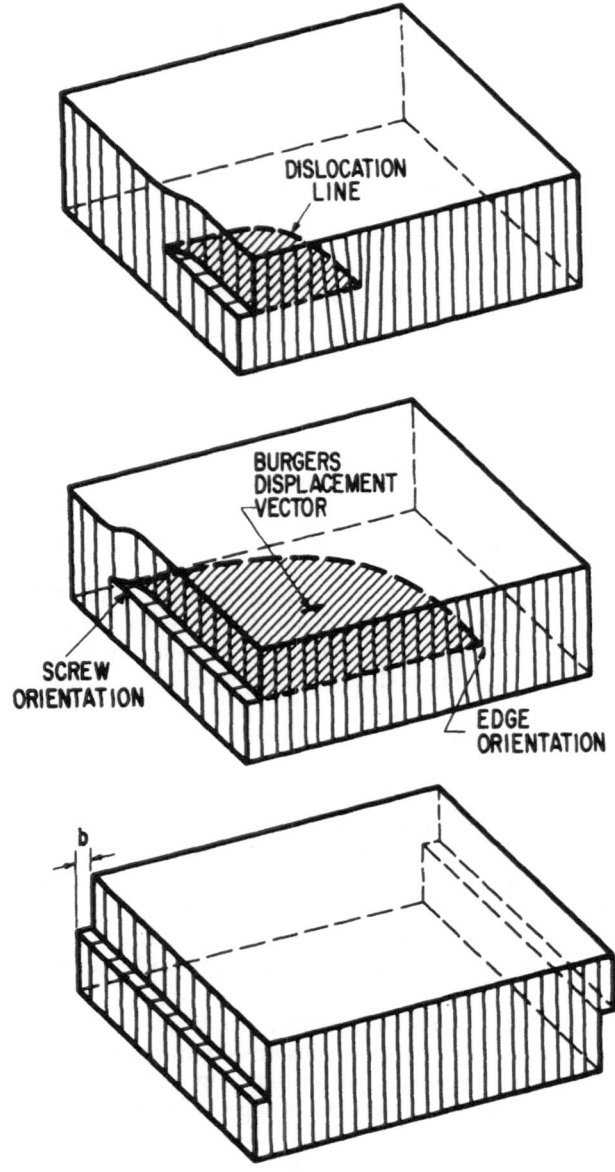

Fig. 1. Spread of translation gliding across a crystal.

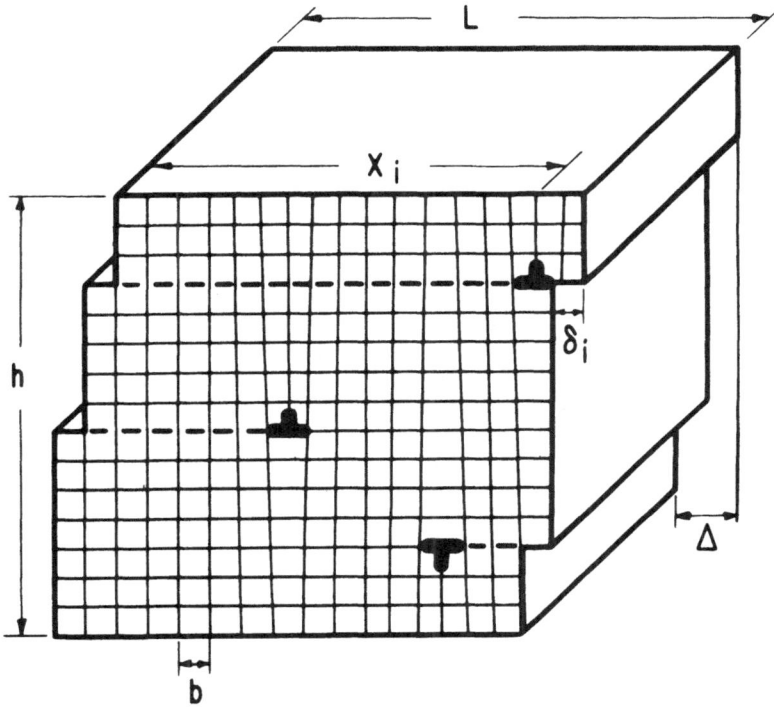

Fig. 2. The strain caused by dislocations.

small region in a crystal, and then gradually propagates through it. The dislocation line, which marks the boundary between the part that has been plastically sheared and the part which has not (boundary between the cross-hatched area and the rest of the glide plane), moves until it has passed entirely through the crystal, whereupon a permanent change of shape has occurred and the dislocation line is gone. Therefore, the only thing that is changed is the shape of the crystal. After the dislocation has passed completely through it, the structure of the crystal is exactly the same as at the beginning. The structure of the crystal is changed, however, during the time that the dislocation line is present within it. Therefore, if the initial dislocation multiplies as it moves through the crystal, the structure changes and strain hardening results.

Next, the flow equations will be derived with the aid of Fig. 2, which shows a few dislocation lines moving through a crystal. They have moved various distances, starting from one side and going across to the other. The macroscopic plastic strain γ is defined as the displacement of the top of the crystal relative to the bottom, divided by the height of the crystal. That is

$$\gamma = \Delta/h$$

But the total displacement of the top is made up of many small displacements associated with individual dislocation lines. So Δ equals the sum of the δ_i's (where i ranges from 1 to n):

$$\Delta = \sum_{1}^{n} \delta_i$$

The individual small displacements depend on how far each dislocation line has moved. For example, consider the dislocation represented by the inverted $\top$ near the top of the crystal. It has passed nearly all the way through the crystal so its δ_i equals one periodic spacing of the crystal, b, which is called the Burgers displacement. When this dislocation began to move from the left in Fig. 2, δ_i was zero. As it moved across to the right-hand side, the displacement increased from zero toward b. Since the displacement is very small compared to the total size of a real crystal, the behavior is linear and the displacement δ_i is simply the fractional distance that the dislocation has moved (that is, x_i is divided by the total width of the crystal) times the total Burgers displacement:

$$\delta_i = (x_i/L)b$$

Then, when the dislocation is just beginning to move $x_i = 0$, so $\delta_i = 0$. When it has traversed the crystal, $x_i = L$ and the displacement is equal to the Burgers displacement. Summing over the δ_i's yields the total displacement.

It is difficult, if not impossible, to measure the x_i's for all of the dislocations in a crystal because there may be millions of them and no really precise method for watching them at every instant of their motion exists. Therefore, we replace x_i by its average. Then the displacement becomes equal to the average distance moved times the number of moving dislocations:

$$\Delta = bn\bar{x}/L$$

and the strain becomes

$$\gamma = n\bar{x}b/hL$$

Since dislocations are lines, it is convenient to speak of the flux ρ of dislocations that pass through a unit of area. This is the area density of dislocations:

$$\rho = n/hL$$

The expression for the plastic strain then becomes

$$\gamma = b\rho\overline{x}$$

and, as might be expected, the strain increases with the number of moving dislocations, the average distance that each moves, and the size of the Burgers displacement.

In a flow process, the total strain has much less significance than the strain rate, so an expression for this is needed. Another reason for studying the strain rate instead of the strain is that the difficulty of measuring all the x_i's also makes it difficult to determine experimentally $\overline{x}$, the average distance moved. The expression for the strain rate is

$$\gamma = b\rho\overline{v}$$

which contains the average velocity, $\overline{v}$, instead of $\overline{x}$, and this can be measured readily.

For properties like the ductility of materials, not only the strain rate is important, but also the response of the material to transient stresses. This is described by the second derivative of the strain with respect to time, or the rate of change of the strain rate:

$$\ddot{\gamma} = b(\rho\overline{a} + \overline{v}\dot{\rho})$$

where a is the average dislocation acceleration, and $\dot{\rho}$ is the rate of change of dislocation density.

This completes specification of the material parameters that determine plastic mechanical behavior. Now their values can be compared for metallic and ceramic crystals. Some of them are fixed in such a way that they cannot be varied and hence are of little interest to the investigator. For example, the Burgers displacement is known from X-ray crystallography, and for a wide variety of crystals it does not change much from one structure to another. Although atomic radii vary quite a bit from one crystal structure to another, the minimum translation distance in a fairly simple crystal (this does not apply to molecular crystals) does not. Furthermore, there is no way of changing the Burgers displacement once the material is given.

The average acceleration of a dislocation is determined mainly by the density of the material because it depends on the effective mass of the dislocation. Again the investigator has no control over this property once the material is given.

The dislocation density at a given instant is already determined at the start of the stress–strain curve, and this initial dislocation density is determined by the method that has been used to fabricate the crystal. If the crystal is made very carefully, it can be very small; but if the crystal is made crudely, it can be very large. In either case it is determined and nothing can be done to vary it independently. If the material

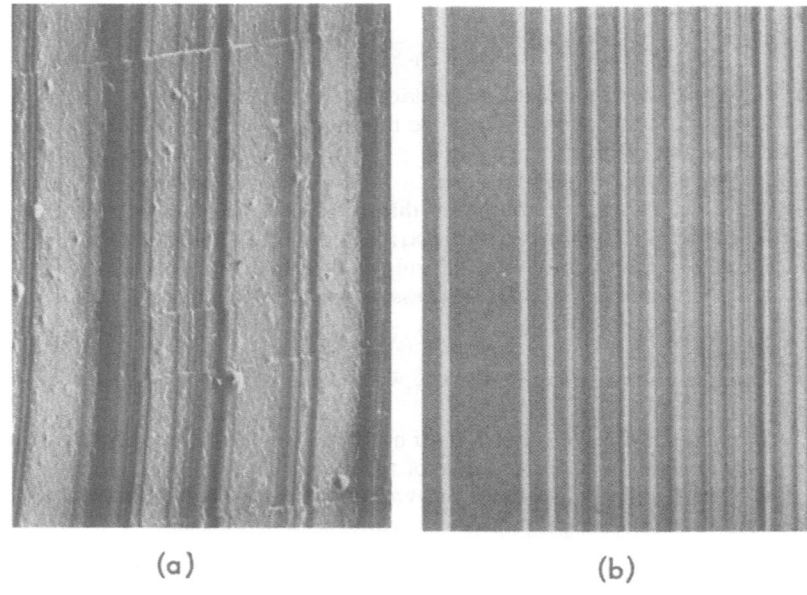

(a)   (b)

Fig. 3. Comparison of glide lines in a metal and an oxide crystal; (a) copper
(after J. T. Fourie), (b) magnesium oxide (after M. L. Kronberg).

is examined at a given instant in time, the dislocation density can be
measured and numerically specified, but not controlled.

The remaining parameters are the average dislocation velocity in
the material, and the rate at which the dislocation density changes.
Velocity is the controllable parameter that determines the strain rate,
and the multiplication rate together with the velocity determines the
transient response.

SIMILARITIES OF METALLIC AND CERAMIC CRYSTALS

The static properties of crystal dislocations are largely determined
by the elastic modulus and the Burgers displacement. Therefore, there
is not much difference in the static properties between metals and ceramics because these physical properties are comparable for them. As an
example, we see in Fig. 3 that the glide lines in a metal (a) (represented
by brass) and a ceramic crystal (b) (magnesium oxide) look much the
same.

Next, Fig. 4 compares the structures of two crystals after a certain amount of plastic bending followed by annealing. Again the gross features are much the same. In this case, a zinc crystal (a) that has been bent and then annealed to polygonize it is compared with an aluminum oxide crystal (b) that has been bent and annealed. The polygon boundaries look much the same for the two substances.

In Fig. 5 we compare the etching of dislocations in a metal and a ceramic crystal. Part (a) shows the etch pits in copper, and the photograph shows the positions of a dislocation that was moved between two etching treatments. In this way, the motion of an individual dislocation line can be observed in copper. In part (b), a very similar thing may be seen in a calcite crystal (calcium carbonate). Again, although the details of the etch pits are different, in their general aspects they are much the same.

In electron microscopic observations as compared in Fig. 6, the similarity between metal and ceramic crystals still prevails. In this figure, part (a) shows dislocation lines seen in a thin film of stainless steel by electron microscopy, and part (b) shows the same thing in a thin film of magnesium oxide.

The similarities of the last few figures occur because all the characteristics that have been compared depend primarily on the Burgers displacement and the elastic modulus of the material. This is also the reason that the gross aspects of mechanical behavior as revealed by stress–strain curves do not show any striking differences. Figure 7a shows a typical stress–strain curve for a steel specimen consisting of an initial elastic portion, then an upper yield point, followed by a lower yield point, and then the onset of strain hardening. The same thing for a lithium fluoride crystal is shown in Fig. 7b and the curve has the same features as the one for steel. Even the stress levels need not be greatly different.

Finally, if we look at gross aspects of the geometry of plastic deformation in a set of structurally similar, but chemically different, ceramic crystals we don't see any qualitative differences. Figure 8 compares the glide bands that form in a variety of diatomic crystals having ionic binding, and all with the rock salt structure. Glide bands are shown for ordinary rock salt (sodium chloride), then for lithium fluoride, and finally for magnesium oxide. The photographs of lithium fluoride show the difference between ones that form at low stress levels (280 g/mm^2) and at higher stress levels (940 g/mm^2). It may be seen that the bands in soft LiF have similar appearance to those in NaC1, and the bands in hard LiF approach the appearance of the bands in MgO. Thus the structures of the glide bands are not sensitive to the chemical natures of these crystals, but only to their hardnesses.

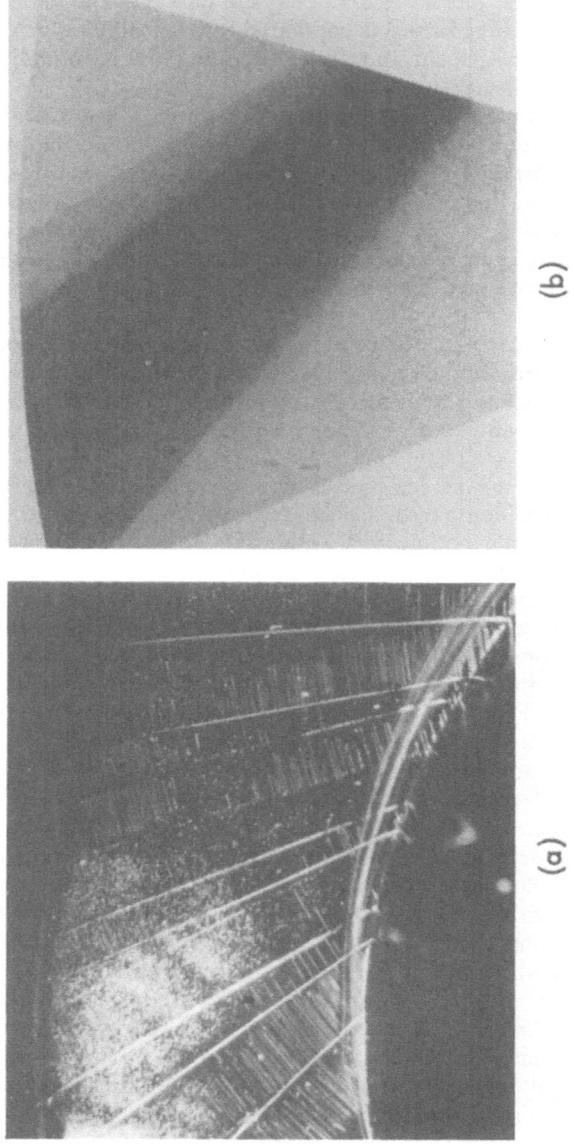

Fig. 4. Comparison of polygonization in a metal and an oxide crystal; (a) zinc, (b) aluminum oxide (after M. L. Kronberg).

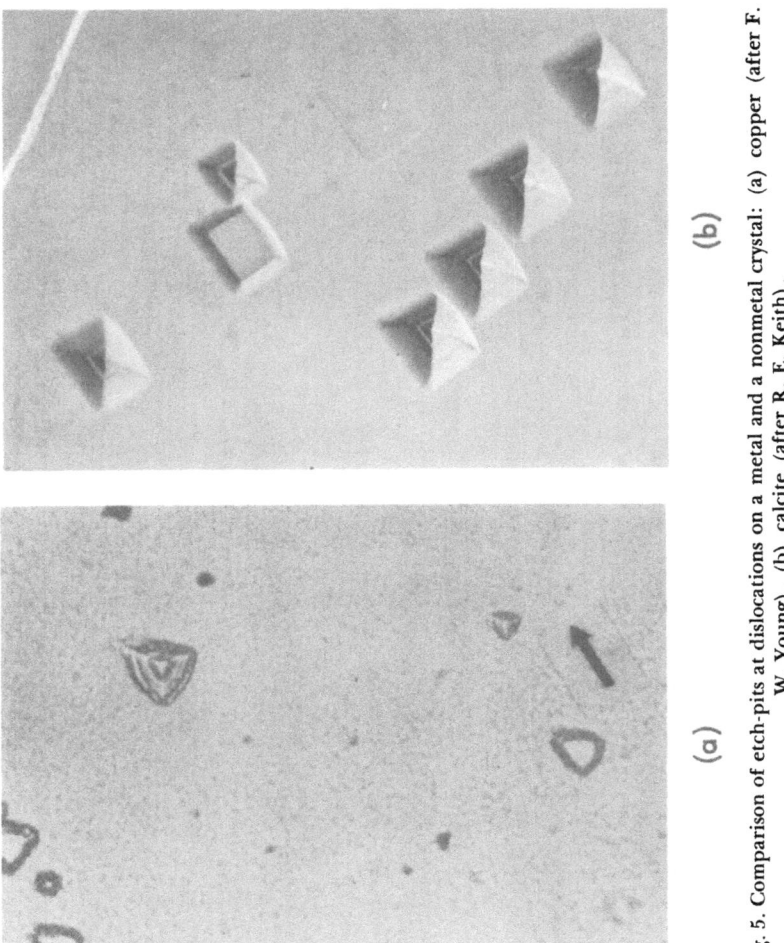

(a)

(b)

Fig. 5. Comparison of etch-pits at dislocations on a metal and a nonmetal crystal: (a) copper (after F. W. Young), (b) calcite (after R. E. Keith).

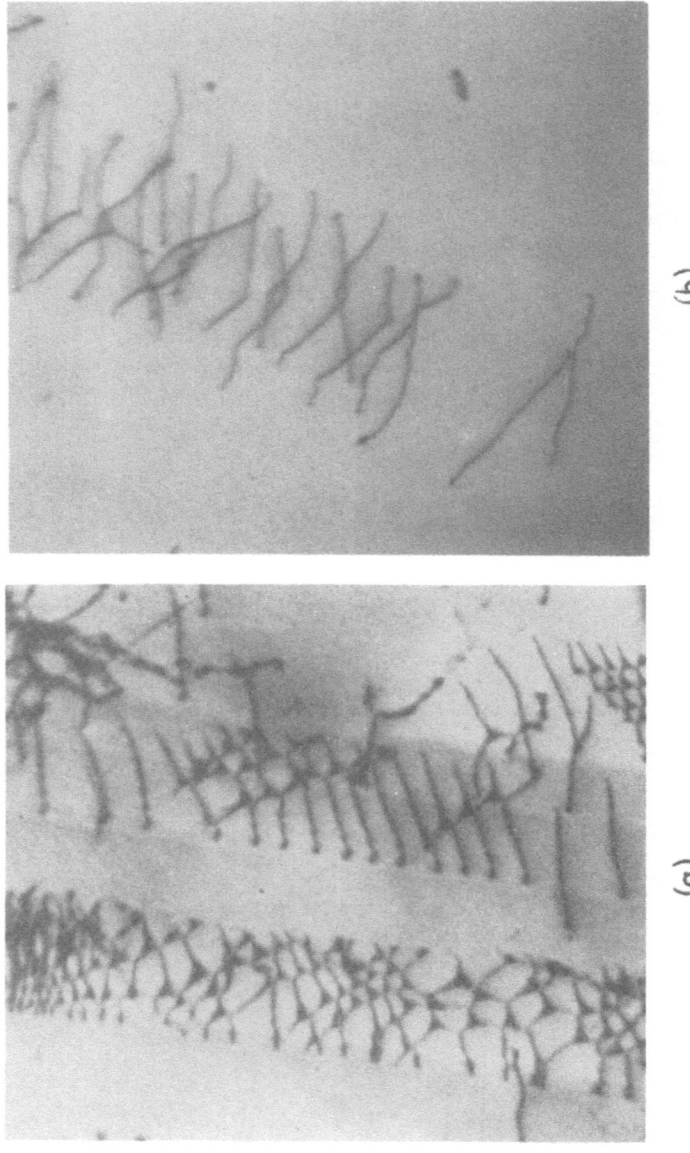

(a)   (b)

Fig. 6. Comparison of electron transmission micrographs of a metal and an oxide crystal; (a) stainless steel (after Hirsch, Partridge, and Segall), (b) magnesium oxide (after Groves, Kelly, and Washburn).

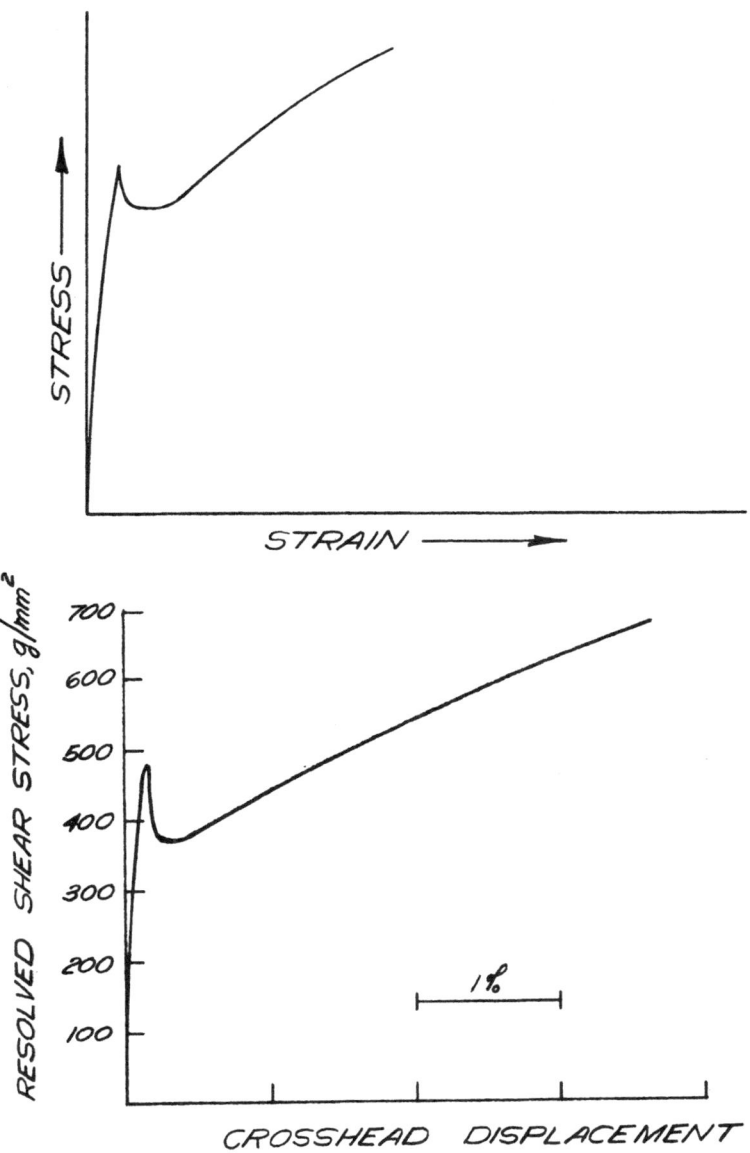

Fig. 7. Comparison of stress–strain curves for metallic and ionic crystals: (a) steel (after D. L. McLean) , (b) lithium fluoride (after W. G. Johnston) .

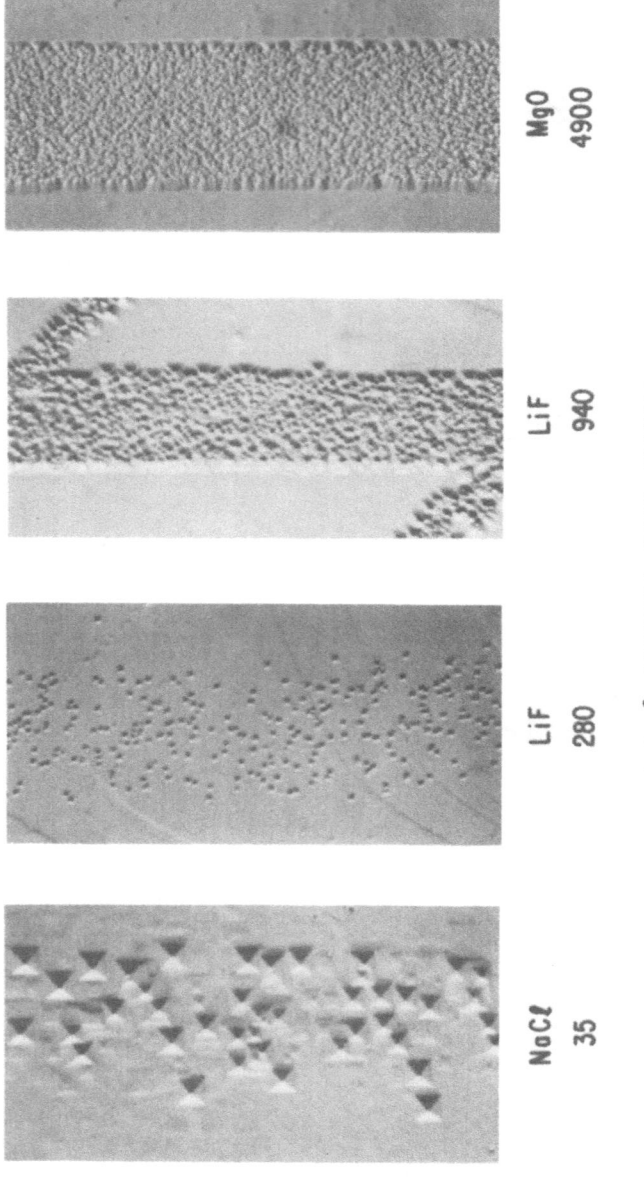

Fig. 8. Comparison of the etched structures of glide bands in ionic crystals (after W. G. Johnston).

SPECIAL STRUCTURAL FEATURES OF CERAMIC CRYSTALS

In contrast to the discussion of the previous section, real differences between metals and ceramics appear when dislocation velocities and dislocation multiplication rates are compared. These effects are related to the structure at the core of a dislocation and Fig. 9 is a schematic diagram that illustrates how the core structure can have a strong influence, even though the effects are rather subtle because most of the dislocation's energy resides outside of the core region.

As a dislocation moves along at various velocities, the energy within the core doesn't change very much with position. From the schematic figure, we can see why this is so. At the top is a drawing of a perfect crystal; just below an edge dislocation has been put into it and arranged so that the extra half plane of atoms is placed symmetrically with respect to the lower part of the crystal. For this arrangement, all of the forces between atoms are balanced on either side of the midplane of symmetry. Hence, there is no net force on the center of the dislocation. Next, the third drawing shows the dislocation moved over to another symmetric position where the bottom half of the crystal becomes symmetric with respect to a plane in the top half, and again everything is balanced so there is zero force on the dislocation.

When the dislocation lies between the symmetry positions, as in the fourth drawing, some imbalance does exist but it is quite small because all the forces are still nearly balanced. Therefore, the overall resistance of the crystal to dislocation motion is quite small. One way of increasing it very considerably is to put an impurity atom at the core of the dislocation so that an unusually strong bond is formed, as in the fifth drawing of Fig. 9. This causes a marked imbalance of the forces and hence resistance to dislocation motion.

Because of the detailed force-balancing at the center of a dislocation, the behavior is very sensitive to small bonding differences and this is reflected in the mechanical behavior of ceramic crystals. Their crystal structures are more complex than those of metals, so that rather specialized structures result at the cores of dislocations in them. Figure 10 shows this for the rock salt structure.

The plane on which the atoms are most closely packed in the rock salt structure is the {100} plane, but the observed glide plane is a less closely packed plane, the {110} plane. In metals, close-packed planes are always preferred for glide, so this difference in behavior must be associated with ionic binding. In fact, if rock salt types of crystals are studied which are only partially ionic and partially metallic (such as lead sulfide), then the glide plane becomes the close-packed {100} plane. In highly ionic crystals there is considerable preference for the {110} plane, however, as the stress–strain data in Fig. 11 indicate. In this figure, data for lithium fluoride show the stresses required to cause flow on

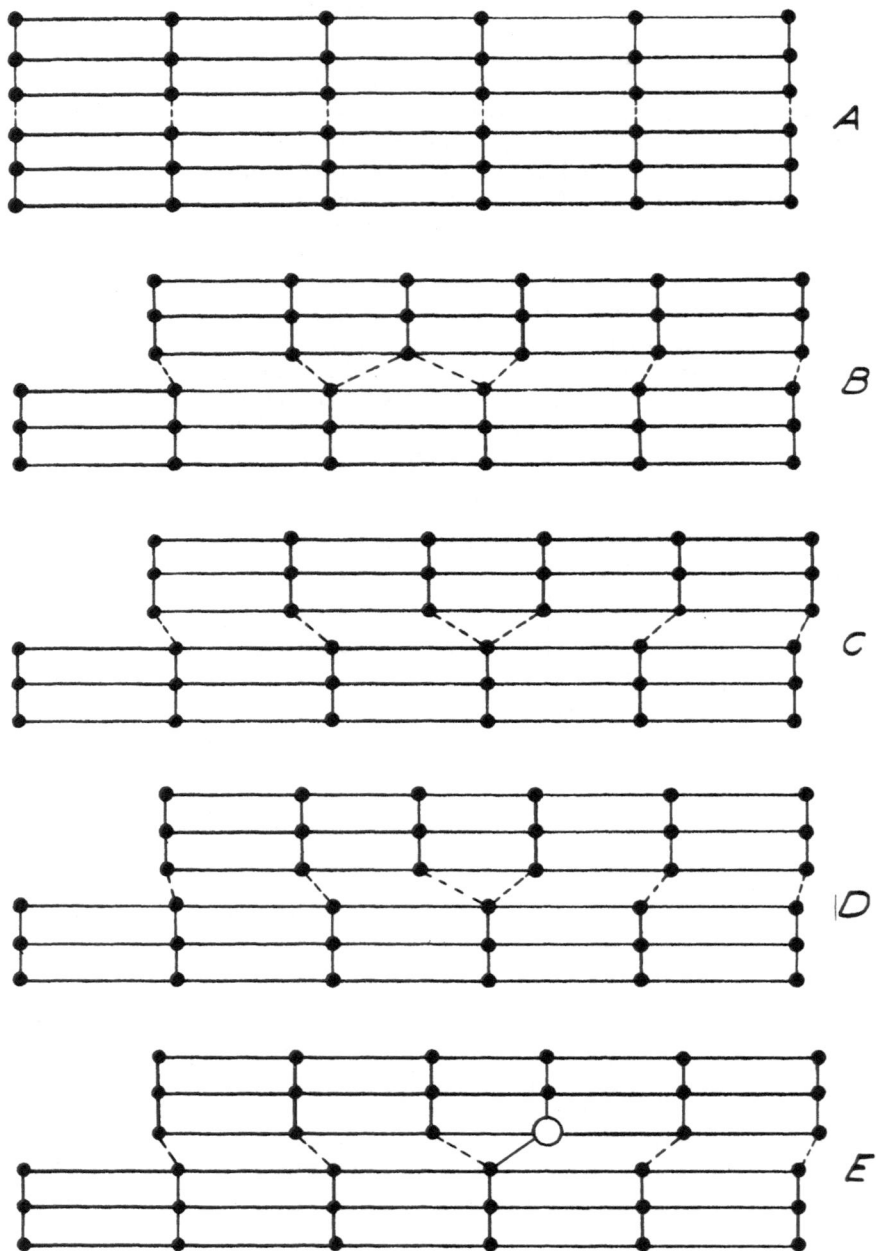

Fig. 9. Schematic drawings illustrating forces at the center of a dislocation line.

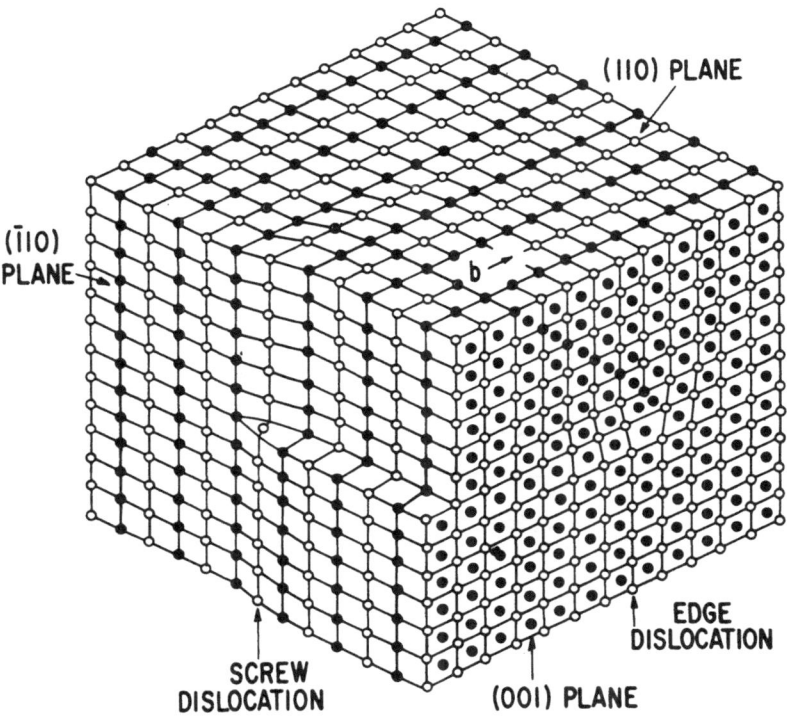

Fig. 10. (110) [1̄10] dislocations in the rock salt crystal structure.

{110} planes together with the stresses required to cause flow in the same glide direction, but on the {100} plane. It may be seen that at low temperatures there is a very large difference between the two stresses.

The reason for preferred {110} glide can be seen if we look at what happens during the motion of a dislocation core with the aid of Fig. 12. The detailed structures are compared as shear occurs on a {110} plane, and on a {100} plane. For shear on the preferred glide plane, when the upper portion gets to the half-glided position, we see that every anion is paired across the glide plane by a cation. Now if the shear is made in the same direction but so as to cause a displacement across the {100} plane, an electrostatic catastrophe developes because the cations at the midglide position are paired across the glide plane by more cations making repulsion (see Fig. 12), and similarly for the anions. Thus, at the midglide position in this case, there is no cohesion across the glide plane. In other words, there has been a very large increase in energy

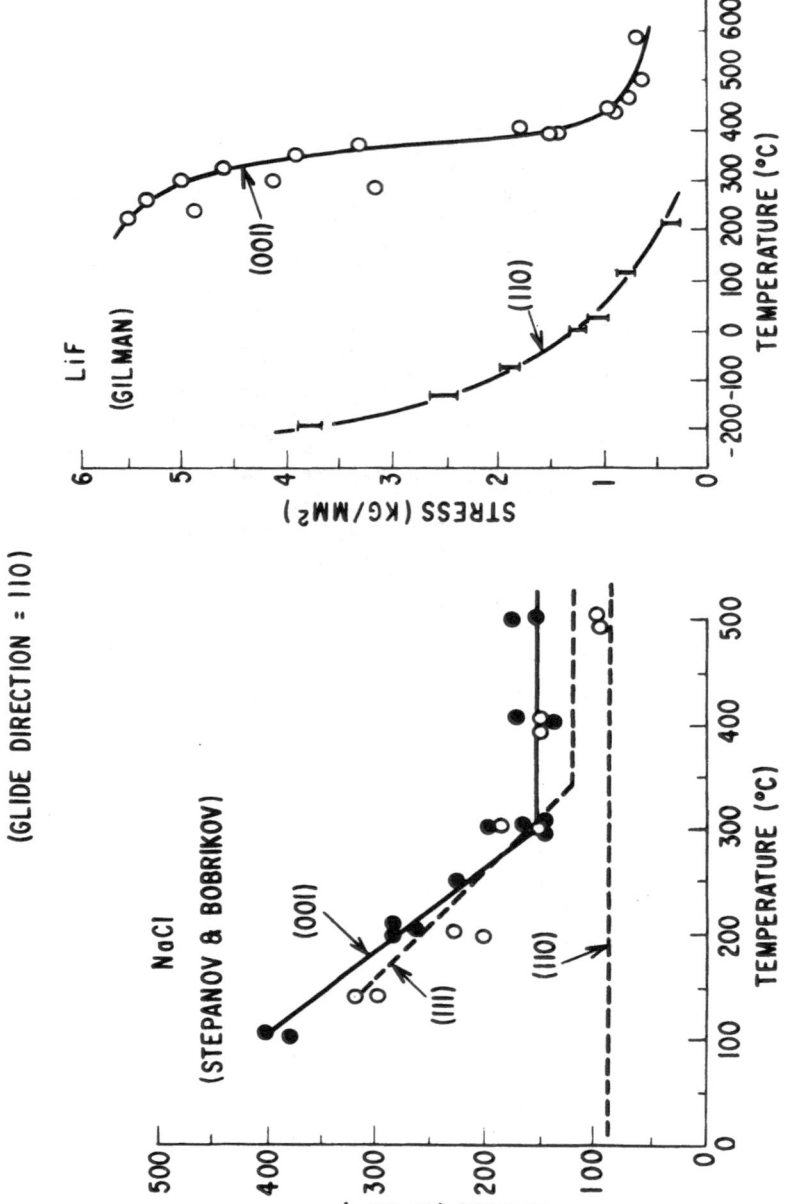

Fig. 11. Comparative stresses for glide on various planes in ionic crystals.

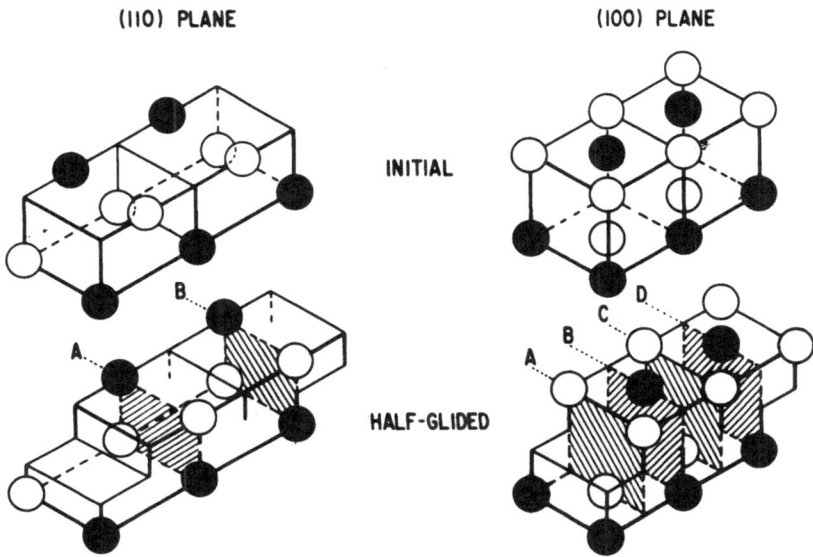

(110) PLANE   (100) PLANE

INITIAL

HALF-GLIDED

Fig. 12. Structures of glide planes in rock salt-type crystals for half-glided positions.

during shearing from the starting position up to the midglide position. Of course, the energy decreases again during shearing to the full-glide position. The electrostatic fault exists only at the midglide position.

Another effect that appears for dislocations in the relatively complex oxide structures is related to symmetry. It is illustrated by Fig. 13, which shows the structure of aluminum oxide. If only the oxygen ions are considered (the large circles), the arrangement is much the same as it is for a close-packed hexagonal metallic structure, and it has hexagonal symmetry. Therefore, there are three equivalent glide directions $<1010>$ in the plane of the figure, and it does not make any difference whether a shear is directed forward along one of these directions or backward. In other words, the gliding process is bidirectional.

If we now consider the structure including the aluminum ions, the symmetry becomes decreased to trigonal, that is, threefold rather than sixfold. Study of the figure will convince the reader that forward motion along one of the directions, A_1, A_2, or A_3, is no longer the same as backward motion. Therefore, if gliding occurs between one layer of oxygen ions in aluminum oxide and an adjacent layer, it will make a difference whether the gliding shear is forward or backward. Because of the reduced symmetry, then, the gliding process becomes unidirectional.

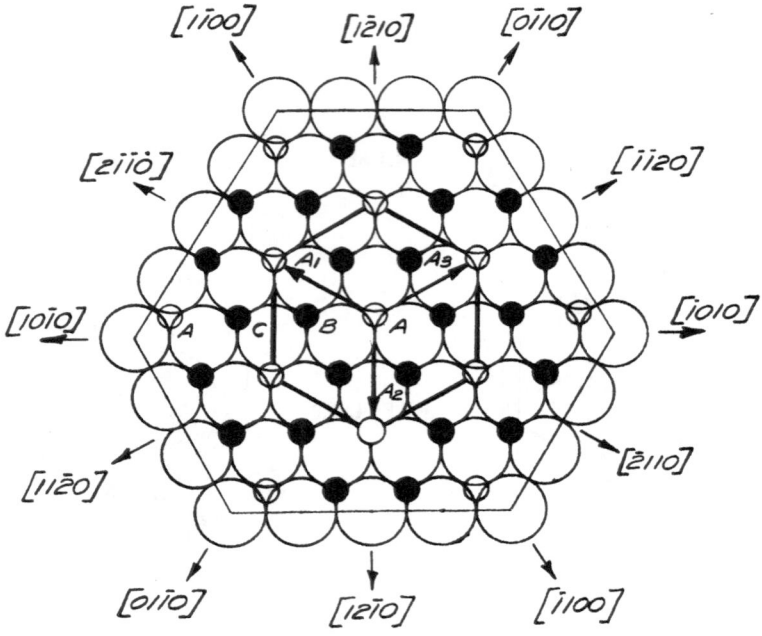

Fig. 13. Structure of aluminum oxide, showing arrangement of aluminum ions and
holes between two layers of oxide ions. Large open circles represent underlying
oxide ions, small open circles represent holes, and small filled circles represent aluminum ions. The upper layer of oxide ions is not shown. Basal hexagonal cell
vectors and directions are shown (after M. L. Kronberg).

There are several crystal structures like aluminum oxide for which
glide is unidirectional. Such a crystal is like a pack of cards with saw-
teeth between the cards so they can slide more easily in one direction
than in the other.

Another complication that occurs in ceramic crystals is decomposition
of unit dislocations into extended dislocations. This also occurs in metal
crystals, but the situations are less complex. Using the aluminum oxide
structure as an example again, the unit Burgers vectors are A_1, A_2, or
A_3 in Fig. 13. It was pointed out by Kronberg some time ago that it is
possible to get less structural disruption during glide by resolving this
unit vector into a series of four smaller displacements as indicated in Fig.
14. In face-centered cubic metals, the corresponding Burgers vector would
be only half as long as in the aluminum oxide structure and it can be
resolved into a series of two steps. Such decompositions are in fact ob-

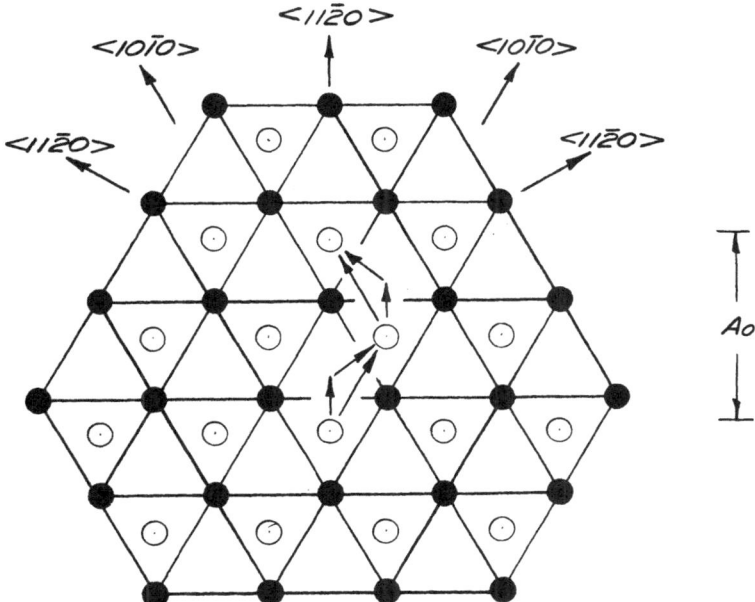

Fig. 14. Illustration of the Burgers vectors of half and quarter partial dislocations —component displacements along $\langle10\bar{1}0\rangle$ and $\langle11\bar{2}0\rangle$ directions, respectively, of the aluminum oxide structure (after M. L. Kronberg) .

served, but ceramic crystals are especially interesting because the more complex decomposition into four partial dislocations can be observed.

No observations of extended dislocations have been made in aluminum oxide but an example in the talc structure (which is rather similar to the aluminum oxide structure) is shown in Fig. 15. Here, instead of the simple lines seen in stainless steel and magnesium oxide (Fig. 6), the dislocations are ribbons consisting of four dislocation lines each.

EFFECTS OF STRUCTURE ON DISLOCATION BEHAVIOR

The various dislocation structures discussed above affect both dislocation velocities and multiplication rates. Dislocations multiply by means of a process known as multiple cross-glide, which is illustrated by Fig. 16. It depends on the ability of screw dislocations to cross-glide; that is, to move off one glide plane onto another. This would be most easy for a cylindrically symmetric screw dislocation because such a dislocation has no definite glide plane and so is free to move with equal ease on any plane. The previous discussion of the structures of disloca-

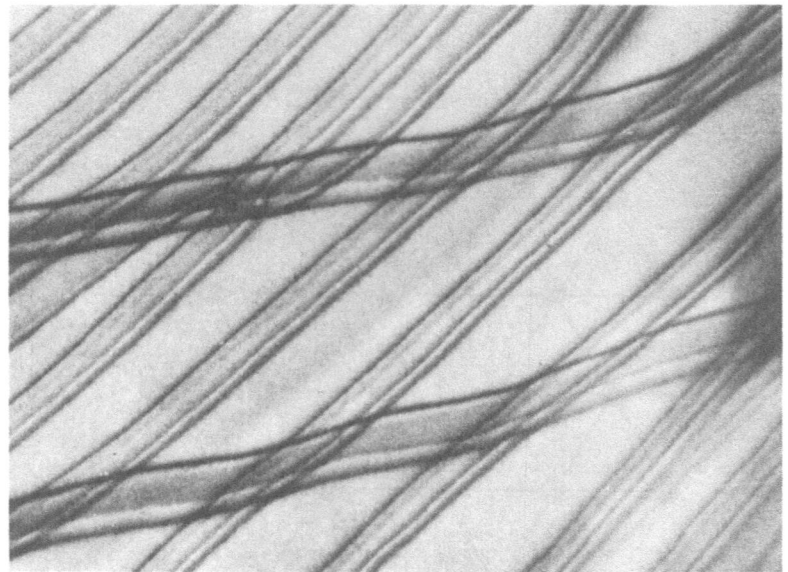

Fig. 15. Ribbonlike extended dislocations in a talc crystal (after S. Amelinckx).

tions in ceramic crystals showed that they cannot be expected to be cylindrically symmetric (sometimes being ribbons) and therefore will experience various degrees of difficulty in cross-gliding and hence in multiplying. This contributes to the brittleness of ceramic crystals at low temperatures. Brittleness also results from low dislocation mobility, but low multiplication rates have an effect. In contrast, metals tend to be relatively isotropic because their binding is not sensitive to structure, so cross-gliding is easy in them, making rapid dislocation multiplication also easy, and thereby allowing them to respond readily to changes in applied stresses. In other words, they have quick transient responses.

Dislocation structure also influences mechanical behavior through its effect on the velocity produced by a given amount of stress. Velocities can be measured, as indicated in Fig. 17, by starting with an individual dislocation half-loop (which connects the two relatively large etch pits in the photograph) and applying a known stress on the loop for a known length of time (so the loop expands and then connects the two more widely separated etch pits in the photograph); repetition of this sequence produced the most widely separated set of pits. Then by measuring the distances that the dislocation lines have moved, and knowing

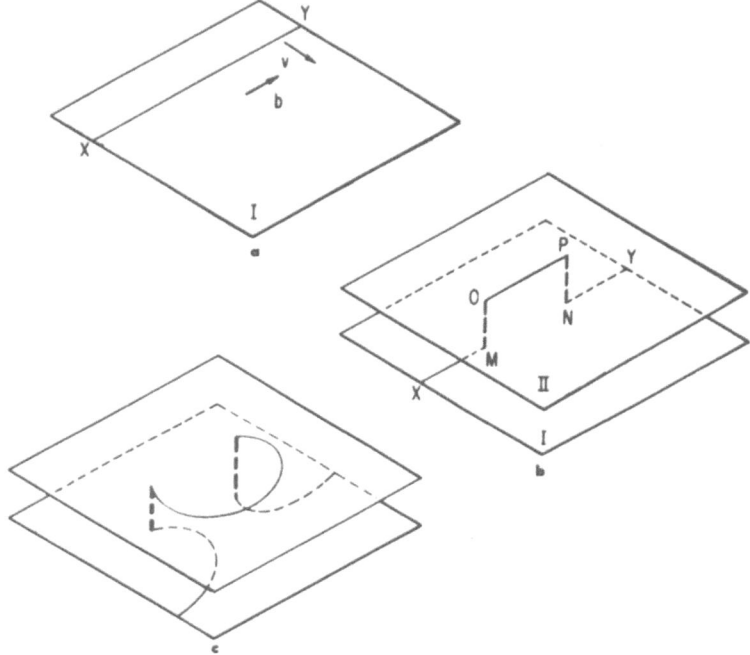

Fig. 16. Schematic drawing of multiplication of a dislocation by cross-glide: (a) moving screw dislocation, (b) segment OP cross-glides to new plane, (c) bowing out of OP to form new loop.

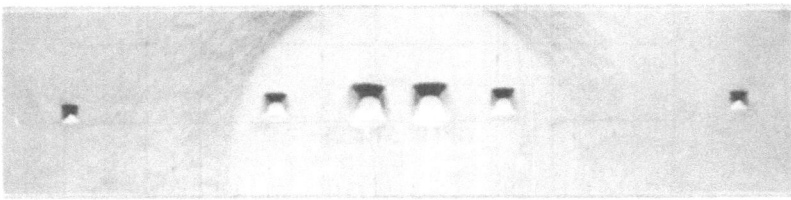

Fig. 17. Expansion of a single dislocation loop in a lithium fluoride crystal. The large pits indicate the original ends of the half-loop; medium pits indicate where the ends were after a stress was applied once. Small pits indicate ends after second stress pulse was applied.

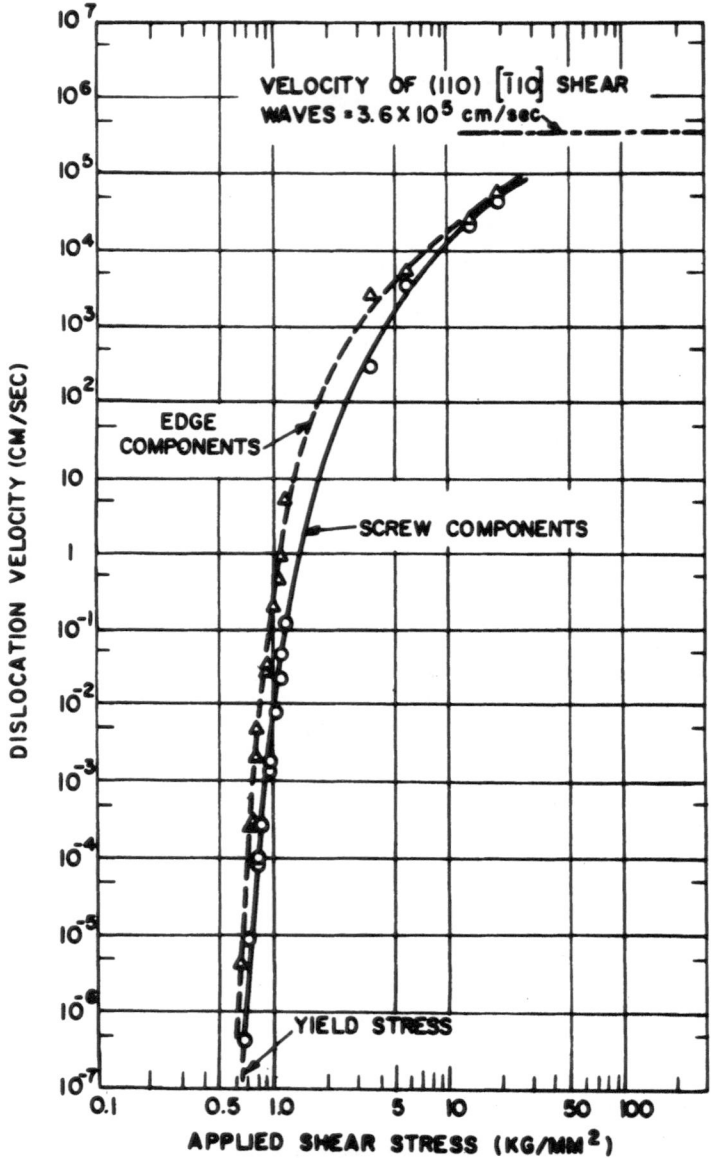

Fig. 18. Effect of applied stresses on dislocation velocities in LiF.

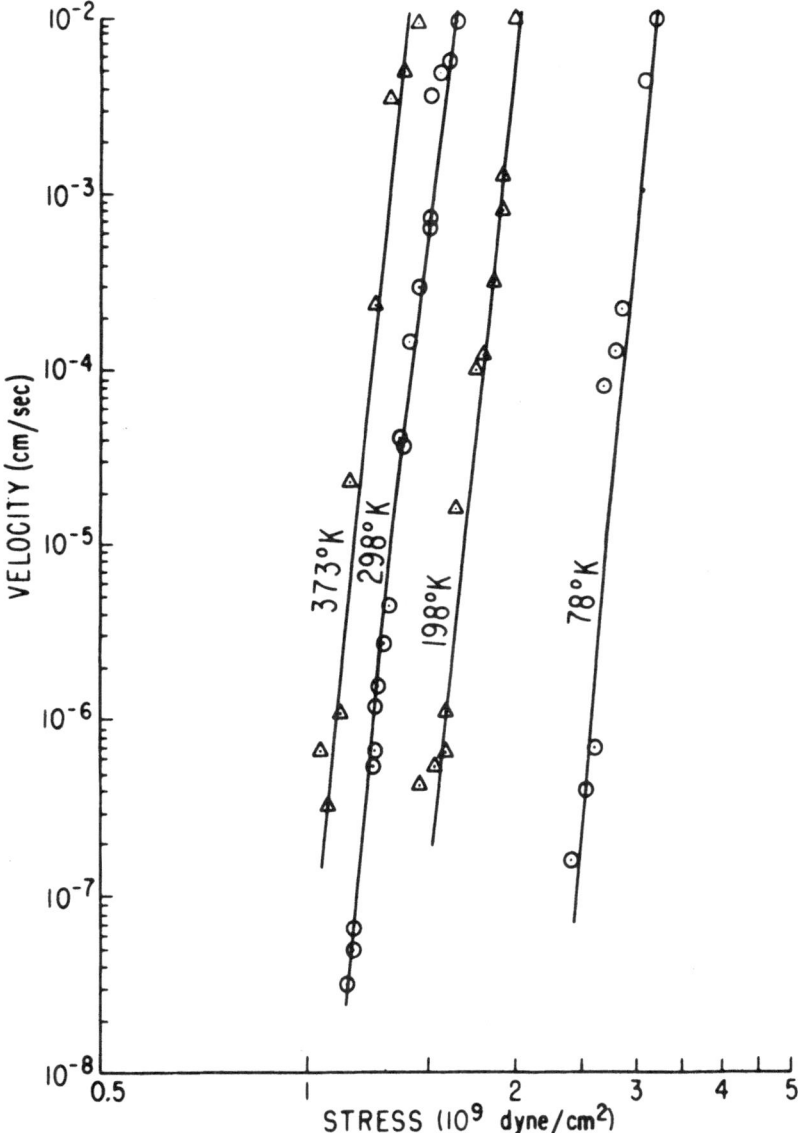

Fig. 19. Edge dislocation motions in Fe–3% Si crystals (after D. L. Stein and J. R. Low) .

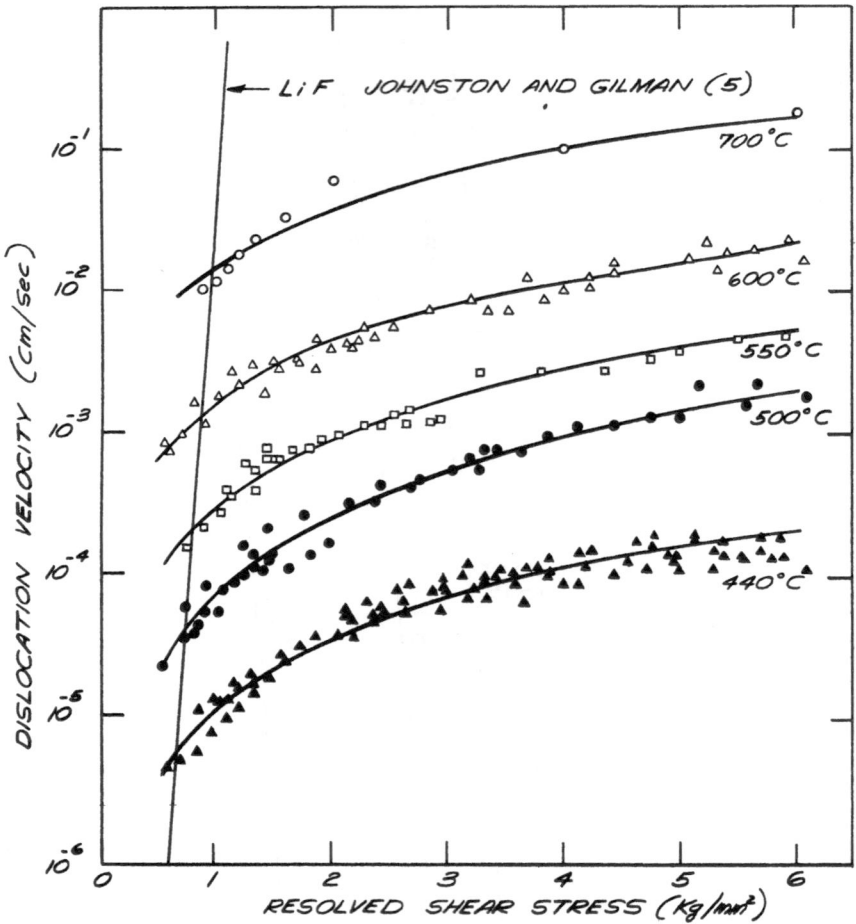

Fig. 20. Effect of stress on dislocation velocities in germanium at various temperatures (after Chaudhuri, Patel, and Rubin).

the time duration of the stress pulse that was applied, one obtains the average velocity.

Data obtained in this way are shown in Fig. 18 for lithium fluoride, which may be considered to be a ceramic crystal since its binding is ionic. In the figure, the logarithm of the dislocation velocity is plotted as a function of the logarithm of the applied stress. There are several features of the data. One is that the edge components of the dislocations

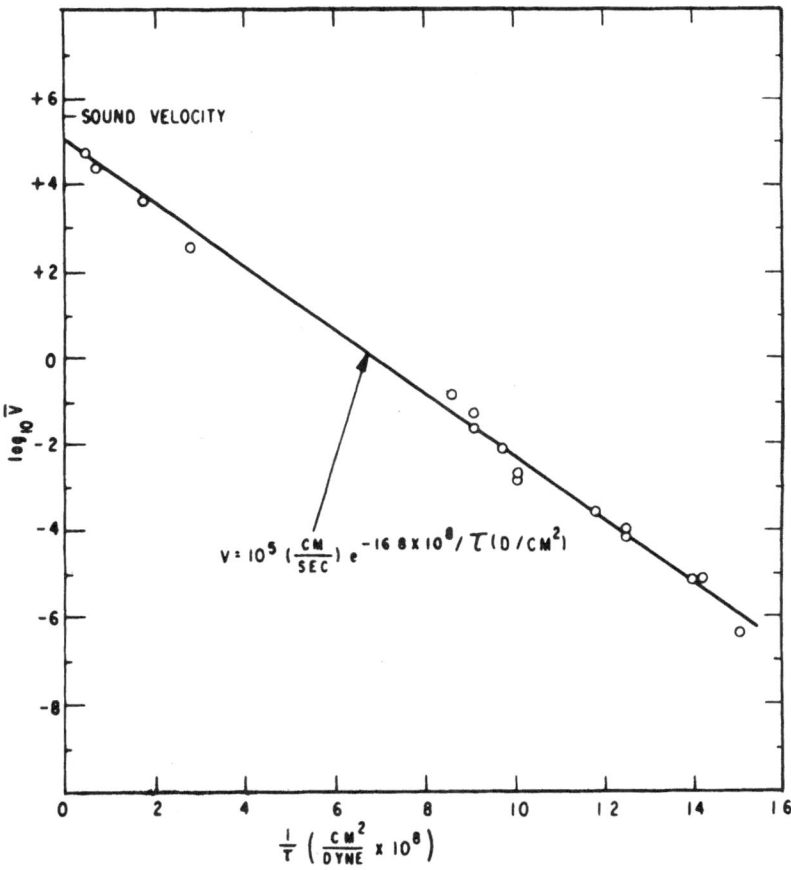

Fig. 21. Plot of logarithm of dislocation velocity vs. reciprocal applied stress (slope of line is a measure of mobility).

move much faster than the screw components. The difference does not appear large on this kind of plot, but it amounts to a factor of about fifty. Second, it is clear that the maximum velocity is sound velocity in the crystal. Third, the velocity is extremely sensitive to the applied stress. This is why plastic flow tends to be relatively strain-rate-insensitive, provided the dislocation velocity is relatively low. The data show that it will become very rate-sensitive as the dislocation velocity becomes very high at high stress levels.

Very similar curves have been obtained for metallic crystals of iron containing 3% silicon (Fig. 19). A new feature of this set of curves is the temperature dependence. Iron is quite sensitive to temperature, so as the temperature decreases, more stress is needed to produce the same velocity. This is the explanation of the increased yield stress of iron as the temperature decreases.

Rather different velocity curves (Fig. 20) have been obtained for germanium, which resembles ceramic crystals with covalent binding. Here the slopes of the velocity vs. shear stress curves are relatively small.

From Figs. 18, 19, and 20 it can be seen that the general shapes of the curves are rather similar, but the quantitative features are quite different. The quantitative difference can be defined by making a different sort of plot as shown in Fig. 21, where, instead of plotting the logarithm of the velocity vs. the logarithm of the stress, the logarithm of the velocity is plotted vs. the reciprocal stress, yielding a simple linear relation that can be expressed analytically. The velocity equals some constant times an exponential function of a negative constant divided by the applied stress. The first constant is related to the velocity of sound but not to details of dislocation structure. However, the negative constant is very much dependent on dislocation structure, and defines the "mobilities" of dislocations. It is the slope of the line and therefore is a measure of the rate sensitivity of plastic flow in the material. It varies considerably between covalently, ionically, and metallically bonded crystals.

CONCLUSION

In summary, some of the similarities and differences between the structures of dislocations in metallic and nonmetallic crystals have been considered, especially from the point of view of their effects on mechanical behavior. There are additional very significant differences that affect electrical and magnetic behavior which I have not even mentioned. Thus it seems clear that studies of the effects of dislocations on physical properties of crystals, especially in the more complicated crystal structures, will continue for a long time in the future.

ACKNOWLEDGMENT

Preparation of this manuscript was supported in part by the Advanced Projects Research Agency.

Plastic Deformation in Polymers

S. S. Sternstein

Department of Chemical Engineering, Rensselaer Polytechnic Institute, Troy, New York

INTRODUCTION

The mechanical properties of polymers, and all matter in general, can be treated from two distinct points of view, namely, the continuum point of view and the structural approach. Each approach has distinct advantages and, indeed, they are complementary. However, more often than not, the two are considered mutually exclusive in that little attempt is made to combine the best findings of each approach. In this article, we shall concern ourselves primarily with an elementary understanding of polymer structure and certain aspects of mechanical behavior. In order to avoid misunderstanding, a few general comments will be made and a simple introduction to linear viscoelasticity will be given.

The continuum mechanics approach to mechanical behavior is based upon two distinct classes of equations, one class being the field equations and the other class the so-called constitutive equations. The general field equations represent, in effect, the physical requirements and restraints which all matter must obey. Such laws as conservation of mass and Newton's laws of motion, when appropriately formulated for a continuum, result in a set of partial differential equations which relate the components of the stress tensor, and their position derivatives, to time and space derivatives of displacements of the mass points of which the continuum is composed. These equations are inviolate.

In order to solve, in principle, any arbitrary solid or fluid mechanics boundary-value problem, a constitutive equation is also required. The constitutive equation represents a relationship between the deformation (in general, strain, torsion, and/or rate of strain) of a body and the resultant components of the stress tensor. This relationship is not the same for all materials and, in general, will be peculiar to a particular material. The effects of temperature, pressure, and thermal and mechanical history of the sample are incorporated into the constitutive equation. The familiar Newton's law of viscosity and Hooke's law are examples of simple constitutive equations.

Recently, significant advances in the generalized requirements of constitutive equations have been reported [1]. For example, dislocations and imperfections are no longer excluded from the domain of rational mechanics. The point to be emphasized is that fundamental requirements are now being formulated for constitutive equations, and any attempt to formulate a "molecular" equation must satisfy these requirements.

Due to the complexity of generalized mechanics problems, it is advantageous for the materials scientist or engineer to limit his observations to a single mode of deformation, e.g., hydrostatic compression, simple shear, or uniaxial tension. It then follows that the constitutive equation which develops from such limited observation is not complete. It is by no means a simple task to generalize from the limited equation to the general constitutive equation for the material, and a superposition of equations developed from single-mode responses is only a first approximation to reality. However, until the philosophical aspects of constitutive equation formulation take on a more easily treated mathematical form, we must content ourselves with the approximate nature of our results.

ANALYSIS

A relatively simple mathematical approach which has been used extensively to analyze the mechanical response of amorphous and rubberlike polymers is the method of retardation-time spectra [2—4]. A retardation time may be thought of as the ratio of the viscous (dissipative) force acting on a flow unit to the elastic (conservative) response of the unit to a given rate of deformation. A flow unit may be several segments of a polymer chain, a side chain, a crystalline region, a domain of an amorphous region, or indeed, any ensemble of atoms, molecules, or structural domains which respond cooperatively to a macroscopic deformation of the body. Due to the complexity of a polymeric structure, it is not difficult to conclude that a distribution of retardation times, not a single retardation time, is necessary to characterize the behavior of a bulk polymer.

The retardation-time distribution, or spectrum, of a polymer can be used to predict the creep compliance from the following equation:

$$J_c(t) = J_g + \int_{\tau_a}^{\tau_b} J(\tau)(1 - e^{-t/\tau})d\tau + t/\eta \qquad (1)$$

where $J_c(t)$ is the creep compliance, J_g is the glasslike compliance, $J(\tau)d\tau$ represents the number of flow units having retardation times between τ and $\tau + d\tau$, t is time, and η is the flow viscosity of the polymer, which may be a function of stress, rate of strain, and temperature. In practice, equation (1) is used to obtain the retardation-time spectrum from the experimentally measured creep compliance [2].

Inspection of equation (1) reveals that the retarded elastic response of a flow unit is proportional to the factor $(1 - e^{-t/\tau})$. It therefore follows that any load application which is performed in time t much less than τ_a will not result in appreciable retarded creep or viscous flow (t/η). Conversely, if the time scale is much larger than τ_b, viscous flow will predominate. Polymers are unique materials in that their retardation times are of the same order of magnitude as normally encountered experimental time scales. Thus, they appear neither viscous nor elastic, but viscoelastic.

Interpretation of retardation spectra and creep curves in terms of molecular parameters has been given in detail elsewhere [2, 3], where it has been shown that equation (1) is a valid representation of creep in amorphous polymers. The glasslike compliance is thought to be representative of elastic, nonretarded deformations, namely, the stretching and bending of covalent bonds, and the perturbation of molecular force fields. The viscous-flow term represents the "plastic deformation" of amorphous polymers, and is indicative of the translation of one polymer molecule relative to another, the motion occurring by means of segmental diffusion. Such an interpretation has been proven correct by showing that increased chemical cross-linking of a polymer increases the viscosity η. When sufficient crosslinks have been introduced to form a three-dimensional polymer network, no viscous flow occurs and η is essentially infinite.

The retarded elastic term is also a manifestation of segmental diffusion which occurs on the application of a stress. However, the retarded elasticity brings into play a thermodynamic restoring force which does not exist for pure plastic deformation. It is the nature of this restoring force and the concept involved, namely, the conformational entropy theory, which is of paramount importance to any study of amorphous, highly plasticized, or elastomeric polymers.

It is convenient to introduce the concept of conformational entropy by first considering the analysis of mechanical deformation by means of classical thermodynamics [5—8]. If a reversible, isothermal deformation of a polymer is considered, the first law of thermodynamics may be written as

$$dE = TdS - (pdV + fdl) \qquad (2)$$

where f is a mechanical force, l is a measure of deformation (in uniaxial tension, l is simply the length of the sample), and the other symbols have their usual meanings. If the deformation process occurs at constant volume (simple shear), or if the material is nearly incompressible (rubber), then we may write equation (2) as follows:

$$f = f_E + f_S \qquad (3)$$

where f_E is the energic force component, f_S is the entropic force component, and these quantities are defined by the equations

$$f_E \equiv \left(\frac{\partial E}{\partial l}\right)_{T,V} \tag{4}$$

$$f_S \equiv -T\left(\frac{\partial S}{\partial l}\right)_{T,V} \tag{5}$$

Furthermore, it can be shown [5] that equation (5) may be obtained in terms of conveniently measured experimental quantities:

$$f_S \equiv -T\left(\frac{\partial S}{\partial l}\right)_{T,V} = T\left(\frac{\partial f}{\partial T}\right)_{l,V} \tag{6}$$

From equation (6), it is noted that the force–temperature coefficient, at constant deformation, is sufficient to determine the entropic force component. Substitution of equation (6) into equation (3) gives

$$f_E = f - T\left(\frac{\partial f}{\partial T}\right)_{l,V} \tag{7}$$

Thus the energic and entropic components of force can be determined experimentally for a given material, provided that an equilibrium relationship among force, temperature, and length can be obtained. In practice, this may require long time intervals between measurements. In many cases, such an equilibrium relationship is nonexistent, since the material is in a state of metastable equilibrium and is kinetically hindered from reaching an equilibrium state, e.g., a glasslike material below its glass transition temperature.

Such cases notwithstanding, the application of equation (3) can provide fundamental insight into the mechanisms of mechanical deformation. For metals, glasses, ceramics, and crystalline and amorphous polymers below their glass transition temperature it is found that f_E predominates, and the deformations involved can be related to the cumulative effects of bond-angle deformation and strain and the pertubation of molecular force fields. Conversely, for elastomers, glasslike polymers above their glass transition temperature, and highly plasticized polymers (e.g., wet cellophane) the entropic force begins to predominate. Perhaps of more significance is the fact that metals and ceramics generally have negative entropic forces, while rubberlike materials have positive entropic forces. Associated with these positive entropic forces are low moduli of elasticity and very large extensions to break.

From a molecular point of view, the energic forces arise from stretching of covalent bonds, distortion of bond angles, and perturbations of

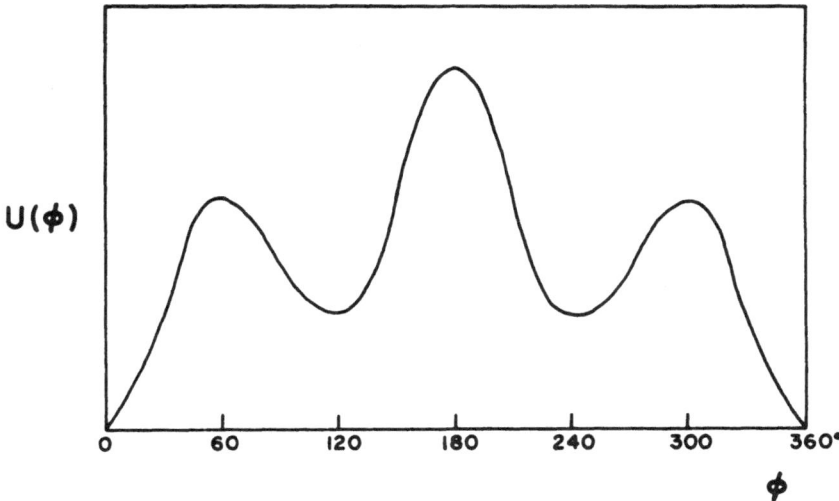

Fig. 1. Potential energy as a function of angle of rotation in polyethylene.

intermolecular force fields, the latter including changes in van der Waals forces if a constant volume deformation is not carried out. The entropic force may be due to an order–disorder transformation as a result of deformation, the increase in entropy which occurs when a covalent bond is stretched, or perhaps the adsorption of a chemical from the environment of the sample (e.g., the adsorption of water vapor by rayon upon stretching).

High extensibility, low modulus of elasticity, and a large decrease in entropy with stretching, which results in a positive entropic force [see equation (6)], are all characteristic of elastomeric behavior. The entropy changes which occur are referred to as conformational entropy changes and result from an ordering of macromolecules upon stretching. A brief review of the basic molecular phenomena which give rise to these entropy changes is given below. An excellent description of these concepts is given in [5].

Atoms linked together by single covalent bonds so that they form a chain molecule have the ability to rotate with respect to each other about an axis which is drawn between the centers of the two adjacent atoms. The rotational motion is resisted by a rotational potential energy function which represents the energy of the rotator as a function of the rotational angle. In any large ensemble of possible rotators, e.g., the carbon-to-carbon covalent bonds in the backbone of a vinyl series polymer, the number of units with energy sufficient to rotate is governed by a Boltz-

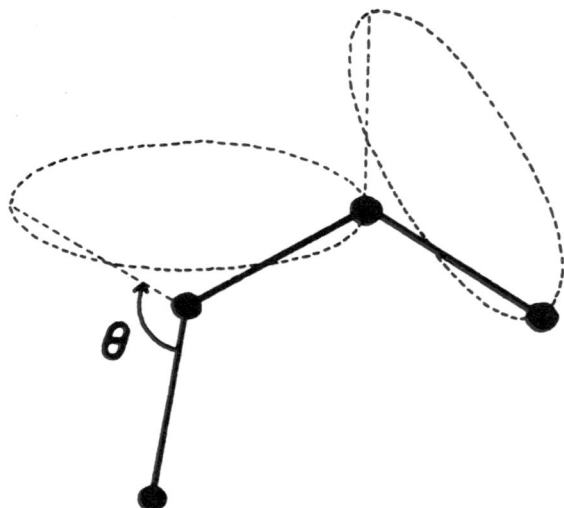

Fig. 2. A freely rotating polymer chain with fixed bond angles. Three segments of the chain are shown.

mann factor, while those units which do not rotate execute torsional oscillations about some average position. The shape of the potential energy curve for rotation in polyethylene is shown in Fig. 1. Various potential barriers to rotation have been summarized in detail [10].

The form of the potential-barrier curve will vary with the atoms under consideration, and most important for our purpose, will vary with the substituent groups attached to the rotators under consideration. For example, if a carbon backbone macromolecule is being considered, then the rotational ability will vary with the types of atoms attached to the carbon. Hence the barrier to rotation in polyethylene is different than the barrier for poly-methyl-methacrylate. Furthermore, because the "rotational freedom" is governed by a Boltzmann factor, the ability to rotate will vary with temperature.

If the repeating units of a linear polymer chain are depicted as single lines, and these units are required to make a bond angle θ with respect to each other (e.g., the tetrahedral angle for a carbon chain), then the result of free rotation is indicated schematically in Fig. 2. The ability to rotate results in a coiled macromolecule as depicted in Fig. 3, where the probability of a given end-to-end distance is shown in Fig. 4. Upon application of Boltzmann's hypothesis, which states that entropy is proportional to the logarithm of equienergetic microstates, to the polymer molecule depicted in Figs. 3 and 4, it is apparent that mechanical ex-

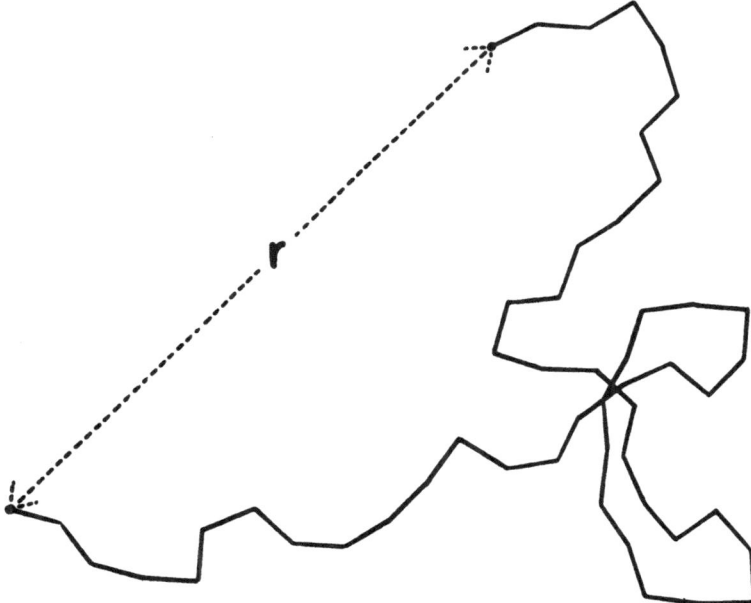

Fig. 3. Random coil conformation of a polymer resulting from internal rotations.

tension of a macroscopic body composed of freely rotating macromolecules results in a lowering of the body's entropy. The decrease in entropy
results in a positive entropic force component as indicated by equation
(5). A complete treatment of conformational entropy is given in [10],
and the application to rubber elasticity problems in [5] and [9].

In real polymeric systems the rotations are hindered not only by
intramolecular barriers but also by intermolecular force fields. Thus, in
order for the polymer chains to have sufficient energy to overcome all
barriers to rotation, the temperature must be above a certain critical
value. Below this critical temperature, which is referred to as the glass
transition temperature, insufficient kinetic energy is available to overcome
all barriers to rotation. Since Fig. 4 represents the equilibrium distribution of chain lengths, and temperature will strongly influence the rate
at which the polymer chain can go from one conformation to another,
it is conceivable that a system can be "frozen" in a nonequilibrium distribution of chain lengths. Such a system can be obtained by quickly
freezing a strip of stretched rubber.

A system kinetically hindered from returning to the most probable
distribution of chain lengths is in a state of metastable equilibrium, and

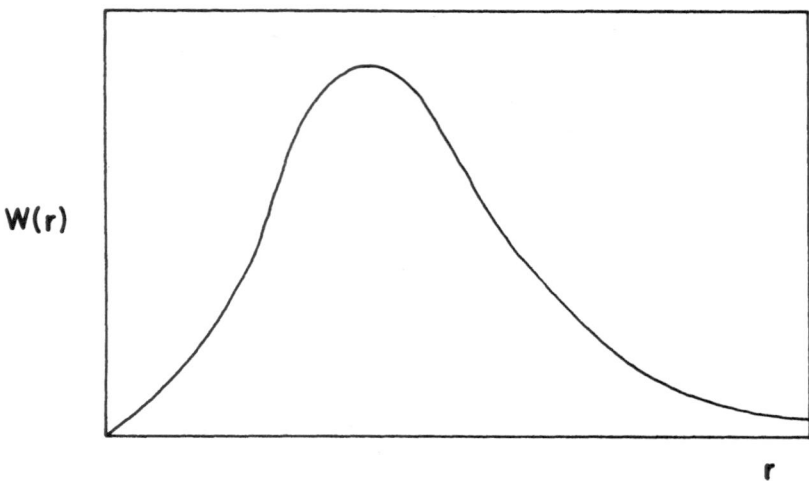

W(r)

r

Fig. 4. Probability of obtaining a given end-to-end distance of the polymer chain depicted in Fig. 3.

it is to be noted that such a system does have a thermodynamic restoring force tending to cause contraction; however, the rate of contraction may be essentially zero. It would be incorrect to say that such a system is plastically deformed since plastic deformation is by definition irreversible.

Attempts to characterize the segmental diffusion concept and describe the motion in linear, amorphous high polymers have been partially successful. However, most treatments are limited to temperatures above the glass transition, or to molten polymer viscosities [11]. For such polymers (e.g., poly-methyl-methacrylate, polystyrene), plastic deformation below the glass transition must involve the same segmental diffusion process as do local rearrangements and changes in conformation. Thus, what may appear to be pure plastic deformation almost certainly contains some amount of kinetically hindered, but thermodynamically recoverable, conformation change. The description of plastic deformation in such a system must be predicated on a fundamental understanding of motion in the glassy state.

The mechanism of plastic deformation in highly crystalline polymers such as nylon and polyethylene is now considered to be closely related to the mode of plastic deformation in metals. Zaukelies [12] has shown that slip planes occur when nylon 66 and nylon 610 are compressed. He argues effectively that the slip occurs within the crystalline regions, and that the slip is due to the motion of dislocations. According to

Zaukelies, the dislocations are due to lattice defects caused by chain ends or chains folding back upon themselves. Furthermore, he has shown that the slip directions as calculated from the crystal structure of nylon are in agreement with his observations.

While dislocation theory can be used to much advantage to describe the deformation of highly crystalline polymers, amorphous polymers appear to be most amenable to treatment by the segmental diffusion concept, for which a complete kinetic description, which includes the effects of stress, temperature, conformational entropy restoring force, and rate of deformation, is not available. Such an analysis would represent a major step forward in our understanding of deformation in polymers.

REFERENCES

1. C. A. Truesdell, Bingham Medalist Address, Society of Rheology, Brown University, August 1963.
2. J. D. Ferry, *Viscoelastic Properties of Polymers,* (Wiley, New York, 1961).
3. A. V. Tobolsky, *Properties and Structure of Polymers,* (Wiley, New York, 1960).
4. T. Alfrey, *Mechanical Properties of High Polymers,* (Interscience, New York, 1948).
5. L. R. G. Treloar, *The Physics of Rubber Elasticity,* 2nd Edition (Clarendon, Oxford, 1958).
6. S. M. Katz, *Textile Research J.* 20:16 (1950).
7. D. R. Elliott and S. A. Lippmann, *J. Appl. Phys.* 16:50 (1945).
8. S. Gabrail and W. Prins, *J. Polymer Sci.* 51:279 (1961).
9. A. Cifferi, *J. Polymer Sci.* 54:149 (1961).
10. M. W. Volkenstein, *Configurational Statistics of Polymeric Chains* (Interscience, New York, 1963).
11. F. Bueche, *Physical Properties of Polymers* (Interscience, New York, 1962).
12. D. A. Zaukelies, *J. Appl. Phys.* 33:2797 (1962).

Bibliography and Index

Bibliography on Theory and Technology of Sintering (137 Annotated References, 1958-1963)

Helen C. Friedemann

New York, N. Y.

INTRODUCTION

Powder metallurgy is growing rapidly. Sintering, the most important step in powder metallurgy processing, is still not completely understood, and is the subject of intensive studies. Some investigations on sintering are described in books, but most are written up in journal articles and reports, and are not always easy to find.

It is the purpose of this bibliography to present a selective review of the work done in sintering in the various parts of the world during the years 1958-1963. The 137 annotated references in this bibliography are taken from the literature in the United States (55), United Kingdom (20), Soviet Union (15), Germany (14), France (7), Japan (7), and Poland (5), as well as Austria, Sweden, Holland, Finland, and Rumania.

The references deal mostly with the still controversial subject of the theories behind the material movement going on during sintering, and to a smaller extent with some more practical aspects of this subject. They cover metallic, ceramic, and plastic materials, in order to show the reader the close interrelationship among these three types of materials.

1. Anon., "Effect of Moisture during Sintering of Tungsten and Molybdenum," *Precision Metal Molding* 19(10):88-89 (Oct. 1961).

   Moisture content of hydrogen atmosphere controls nucleation and grain growth during sintering of W and Mo; increasing moisture content decreases number of nuclei and may lead to single-crystal formation; dew point measurements during sintering; percent recovery of tungsten wire indicated.

2. Accary, A., and R. Caillat, "Study of Mechanism of Reaction Hot Pressing," *J. Am. Ceram. Soc.* 45(7):347-351 (July 1962).

   Refractory uranium mononitride of high density may be prepared by hot pressing mixtures of uranium and uranium sesquinitride at 1000°C; steps of process are densification of metal into continuous uranium matrix and reaction of uranium sesquinitride with matrix; during hot pressing uranium–silicon mixtures react before any densification of uranium occurs; pressing temperature required for given density is then practically same as for unreacted powders and previously formed compound.

3. Accary, A., and J. Dubuisson, "Contribution to the Study of Sintering Accompanied by Chemical Reaction," *Mem. Scient. de la Rev. Met.* 58(11):863-868 (Nov. 1961).

   Investigation of sintered U–Si powder mixtures. Differential thermal analysis showed that, at 470-80°C, the reaction between the two elements was appreciably exothermic, the quantity of heat evolved per unit weight of mixture increasing with the Si content.

4. Alekseeva, F. N., *et al.,* "Role of Microdistortions in the Recrystallization of High-Melting Metal Powder Compacts during Sintering," *Izvest. Akad. Nauk SSSR, Otn. Met. Gorn. Delo,* (Jan.-Feb. 1963) No. 1:97-99 (Translation by H. Brutcher No. 6009).

Study of microdistortions set up during pressing of high-melting metal powders and their role in the recrystallization process taking place during sintering. Experimental procedure. Microstresses from compacting of W, Mo, and Cb powders as function of compacting pressure; most effective pressure range. Conditions for growth of recrystallization grains during sintering of compacts at increasing microstresses of lattice and with decreasing thermal stability. The two variables governing the growth of the recrystallization grains of high- melting metals; recrystallization of Mo, W, and Cb compacts under various conditions of sintering.

5. Alsop, H. J., and J. P. Roberts, "Nonstoichiometry of Zinc Oxide and its Relation to Sintering," *Trans. Faraday Soc,* 55:1386-1393 (1959).

6. Amberg, S, "Studies on Sintering of $MoSi_2-Al_2O_3$ Cermets," *Powder Metallurgy* (1961) No. 8:101-112.

Porosity of hydrogen sintered bodies composed of molybdenum disilicide and alumina has been studied as function of alumina content and grain size; samples were sintered in dissociated ammonia and compared with those sintered in hydrogen; it became apparent that samples sintered in the former underwent particularly severe form of attack when exposed to air at high temperatures; this is attributed to formation, during sintering, of Si_3N_4.

7. Amosov, V. M., and V. V. Dianov, "Pressing and Sintering of Electrolytic Powders of Tantalum and Niobium," *Poroshkovaya Metallurgiya* 1(3):14-19 (1961).

Preparation of ductile Ta and Nb powders, and elimination of impurities at high temperature in vacuum. Dependence of density, strength, electrical resistivity, and amount of open porosity on compacting pressure established. High-temperature sintering of Ta and Nb powder compacts characterized by simultaneous occurrence of another process, refining of the powders. In case of Ta, in temperature range 1000-1600°C, rate of eliminating impurities exceeds rate of sintering, so that amount of open porosity increases. In case of Nb, converse is case. In addition there is sharp increase of density in range 1500-1700°C. With both metals at 2000-2200°C, the amount of open porosity increases, owing to elimination of lower oxides, and thereafter falls sharply.

8. Anderson, P. J., and D. T. Livey, "Physical Methods for Investigating the Properties of Oxide Powders in Relation to Sintering," *Powder Metallurgy* (1961) No. 7: 189-203.

Fine oxide powders may have particle sizes as low as 100 A, and there is evidence to suggest that such particles have different bulk energies from similar large particles. The study of the kinetics of sintering requires a knowledge of the extent of interparticle contacts. The characterization of particle size in the range in question (say less than 500 A) is thus important from both an energetic and a kinetic point of view. Shape of particles will also play a part in determining the rate of sintering. The point of energy conditions as a function of particle size also depends, in one of methods used, on accurate measurement of heat of solution.

9. Andrievskii, R. A., and I. M. Fedorchenko, "Activated Sintering of Iron Powders," *Metallovedenie i Termicheskaya Obrabotka Metallov* (Dec. 1960) No. 12:36-39 Translation by H. Brutcher No. 5531.

Comparative study of relative efficiency of various activating methods in sintering of Fe powders: (1) in dry hydrogen; (2) hydrogen with 2%; (3) with 10% moisture; (4) initial oxidation + dry hydrogen; (5) hydrogen for 7 min, then water for 3 min;

(6) hydrogen + HCl atmosphere; (7) same as 6 but oxidized samples; (8) under powdered alumina + 0.1 NH_4Cl; (9) under powdered alumina + 0.1% NH_4F. Treatments most beneficial for magnetic permeability and coercive force. Effects of other treatments. Nature of favorable effect of activation according to procedure.

10. Bernard, R. G., "Processes Involved in Sintering," *Powder Metallurgy* (1959) No. 3:86-103.

   To arrive at a theory of sintering, it is necessary to select a physical property characteristic of the phenomenon, and to study its variations with temperature and time. Among these are: (1) geometrical—linear dimensions, volume, porosity, shape of voids; (2) mechanical—hardness, tensile strength, modulus of elasticity; (3) electrical and magnetic—electrical conductivity and magnetic susceptibility. Drawbacks and advantages of each. Endeavor to list problems arising: (1) sintering of pure metals without appearance of liquid phase or interaction with atmosphere; (2) sintering of mixtures where there is likelihood that solid solutions will be formed or that chemical reactions will take place; (3) sintering in presence of liquid phase.

11. Birchenall, C. E., "The Mechanisms of Diffusion in Solids," in *Reactivity of Solids: Proc. 4th Intern. Symposium,* Elsevier, Amsterdam (1960), pp. 24-37.

   In most crystalline solids diffusion occurs by defect mechanisms, by the motion of interstitial atoms or ions or vacant lattice sites. When the concentration of defects is controlled by thermal fluctuations in an equilibrium system, the activation energy for diffusion is proportional to the sum of an enthalpy for the formation of the defects and an enthalpy for their motion. The defect concentration may be controlled by composition changes, particularly in ionic crystals. When the number of defects from this source greatly exceeds the number created thermally, the activation energy for diffusion contains only the enthalpy of motion of the existing defects.

12. Blackman, L. C. F., "Lattice Defects and the Sintering of Oxides," *Industrial Chemist* 38(454):620-626 (Dec. 1962); 39(1):23-26 (Jan. 1963).

   Critical review of present state of knowledge of sintering oxides, with more detailed consideration of influence of sintering atmosphere and impurity content, and extent to which experimental evidence on these factors supports a lattice defect sintering model.

13. Braun, H., and K. Sedlatschek, "On the Influence of Small Additions of Nonmetals and Metals on the Sintering, Working, and Mechanical Properties of Tungsten," *J. Less Common Metals* 2(2/4):277-291 (Apr.-Aug. 1960).

   Influence of 0.01-1.0 w/o various elements on sintering behavior, workability and mechanical properties of W: C, B, W_2B, W_3Si_2, Al, Be, Ti, Hf, Th, V, and Cr were used as additives. Changes in as-sintered density and hardness, workability by hot rolling, hardness, transverse rupture strength and ductility of worked alloys determined, and related to concentration of alloying elements.

14. Brett, J., and L. L. Seigle, *Fundamentals of Sintering. V,* U.S. AEC Report SEP-257 (Apr. 1961).

   Experiments to evaluate influence of impurities and atmosphere on activity of grain boundary vacancy sinks were carried out in Ni, Cu, and dezincified brass powder. In absence of extraneous effects, the sintering rates appeared to be surprisingly independent of both the atmosphere and the presence of a surface active impurity, S. The latter indicates that in powder compacts segregated impurities have no pronounced effect on grain boundary vacancy absorption. Compacts of 99.6% pure hydrometallurgically produced Cu powder did not densify in vacuum, due apparently to the entrapment of gas in the pores.

15. Brett, J., and L. L. Seigle, *Fundamentals of Sintering. VI,* U.S. AEC Report SEP-259 (Dec. 1961).

In effort to establish proportion of vacancies transmitted to surface along grain boundaries during sintering, an interferometric study of grain boundary grooves in the neck of twisted 3-wire Cu compacts was carried out. Longitudinal boundaries migrated in the neck during sintering, giving rise to undulations on the neck surface. Transverse boundaries in the neck showed little, if any, enhancement of grooving compared to their extensions away from the neck. There was no indication in the shape or rate of formation of grain boundary grooves that substantial diffusion of vacancies along grain boundaries to the surface occurs. It is concluded that vacancies are eliminated during sintering, primarily by annihilation in the grain boundary.

16. Brett, J., and L. L. Seigle, "Shrinkage of Voids in Copper," *Acta Met.* 11(5):467-474 (May 1963).

From investigation of effect of long sintering at 1050°C in hydrogen or argon on porosity of Cu specimens prepared by dezincification of alpha-brass, it is concluded that intragranular vacancy sinks such as dislocations play a negligible role in void shrinkage during sintering of Cu.

17. Brett, J., *et al, Fundamentals of Sintering. IV,* U.S. AEC Report SEP-254 (June 1959).

The disappearance of porosity during sintering in absence of grain boundaries and free surfaces has been studied in Cu. Large-grained porous Cu specimens were prepared by dezincification of brass. After sintering, metallographic observations of the intragranular porosity at regions remote from grain boundaries and free surfaces reveal no significant densification, even though considerable decrease in the number and increase in the size of voids has occurred. It is concluded that intragranular sinks such as dislocations are not effective in eliminating vacancies during sintering. Also, evidence has been obtained that grain boundary activity and sintering atmosphere are interrelated.

18. Brophy, J. H., *et al,* "The Nickel-Activated Sintering of Tungsten" in *Powder Metallurgy*, ed. by W. Leszynski, Interscience, New York (1961), pp. 113-134.

When fine W powders are coated with a uniform Ni layer of the order of atomic thickness, rapid densification occurs during sintering at unusually low temperatures. The process leads to 92% of theoretical density at 1100°C after 30 min, and is retarded by onset of grain growth. Subsequent slower sintering produces 98% of theoretical density after 16 hr at 1100°C. Theoretical and experimental considerations indicate that nickel-activated sintering of W takes place by the movement of W through or on a thin Ni carrier phase. The activation energy for the sintering process is 68,000 cal/mole.

19. Brophy, J. H., *et al,* "Final Stages of Densification in Nickel–Tungsten Compacts," *Trans. Met. Soc. AIME* 224(4):797-803 (Aug. 1962).

Kinetic analysis made of second stage of 2-stage densification of Ni-activated sintering of W; model of process used; rate-controlling mechanism of second stage, as of first stage, found to be solution of W in Ni carrier phase; kinetic differences between stages explained by spheroidization of pores and initiation of grain growth at beginning of second stage.

20. Brophy, J. H., *et al,* "The Investigation of the Activated Sintering of Tungsten Powder," Final report from Metals Processing Lab., M.I.T. (Feb. 28, 1963); see also ASTIA report AD-266,964 (Sept. 1961).

Low-temperature densification of W can be accelerated by small additions (between 0.1 and 4%) of Pd, Rh, Ru, and Pt in a manner similar to that observed with Ni. This acceleration results from an increase in the rate of grain-boundary diffusion which characteristically dominates the densification of pure W. Ir, in contrast to other Group VIII elements examined, was found to retard densification of W at 1100°C. It is suggested that both activated solid state sintering and liquid-phase sintering are similar in nature.

21. Brophy, J. H., *et al*, "Nickel-Activated Sintering of Plasma-Sprayed Tungsten Deposits," *Trans. Met. Soc. AIME* 227:598-603 (June 1963).
   Technology of Ni-activated sintering of W powder has been successfully applied to the densification of plasma-sprayed W. Ni added by infiltration in a zinc solution followed by evaporation of the solvent. After sintering 1 hr at 1300°C, density 95% of theoretical, and transverse rupture strength of 74,000 psi were obtained. Shrinkage was found to be anisotropic and the mechanism of densification was comparable to that found in the Ni-activated sintering of W powder.

22. Brophy, J. H., *et al*, "Sintering and Strength of Coated and Co-Reduced Nickel-Tungsten Powder," *Trans. Met. Soc. AIME* 221(6):1225-1231 (Dec. 1961).
   Experimental data presented lead to conclusion that W sintering can be activated with minimum amount of Ni if coating process, rather than co-reduction process, is used; mode of powder preparation is significant in determining strength of sintered W + Ni compacts; after grain growth begins, strength is determined by both density and grain size.

23. Broquet, C., *et al*, "Frittage avec apparition de phase liquide," *Mem. Scient. de la Rev. Met.* 60(3):171-176 (Mar. 1963).
   Sintering that involves formation of liquid phase; discussion of experiments concerned with obtaining high density in sintered alloys by forming in solid phase just enough liquid phase to fill pores; method of calculating this amount given; method of sintering powders of Cu–Ni or Cu–Ag alloys and mixtures of Cu and Sn powders at temperatures where some liquid is formed, is described and illustrated by micrographs. Pore formation during homogenization because of normal and seemingly inverse Kirkendall effect is included in discussion.

24. Burke, J. E., "Recrystallization and Sintering in Ceramics," in *Ceramic Fabrication Processes*, ed. by W. D. Kingery, John Wiley, New York (1958), pp. 120-131.
   A diffusion model for matter transport during sintering is superior to a plastic flow model in its ability to explain kinetics of the process and the microstructural changes which occur. Grain boundaries serve as vacancy sinks during sintering; thus the densification and grain growth processes are interrelated.

25. Carlson, R. G., and F. E. Westermann, "Hot Pressing of Lead Spheres," *Planseeber. Pulvermet.* 10 (1/2):15-23 (Apr. 1962).
   To obtain better understanding of basic phenomena in initial stages of hot pressing, antimonial lead spheres were hot pressed in steel die at four different temperatures, and constant pressure, and at four different pressures and constant temperatures. Log–log relationship was found for porosity change vs. time, and time and temperature dependent semilog relationship for porosity vs. pressure, for which empirical equations are given.

26. Cegielski, W., "Untersuchungen des Dichteverlaufes in metallischen Sinterwerkstoffen mit Hilfe von Widerstandmessungen" in *Ber. über die II, Internationale Pulvermet. Tagung in Eisenach*, Akad. Verlag, Berlin (June 1962), pp. 109-119.
   Electrical resistivity measurements are valuable in order to follow up process of sintering. Electrical resistivity is a linear function of density at a later stage of sintering. At earlier stages of sintering, electrical resistivity changes are not due to densification, but to processes occurring at the contact points between the particles.

27. Chang, R., and C. G. Rhodes, "High-Pressure Hot Pressing of Uranium Carbide Powders and Mechanism of Sintering of Refractory Bodies," *J. Am. Ceram. Soc.* 45(8):379-382 (Aug. 1962).

Summary of main mechanisms proposed to account for densification in sintering of powders, when carried out by hot pressing. Authors report on experiments in hot pressing of UC powders at very high pressures (10-46 kbar). Temperatures varied from 500-1500°C. Final density achieved at a given pressure was found to be almost independent of temperature and this suggests that the controlling mechanism is not plastic flow.

28. Clark, F., "Sintering" in *Advanced Techniques in Powder Metallurgy*, Rowman and Littlefield, New York (1963), pp. 80-101.

Theory of bonding, infiltration of solder, liquid-phase sintering; sintering of Cu powder, stainless steel powder, molybdenum powder, binary alloys, nickel carbonyl and ruthenium, metal wires; sintering atmosphere, dimensional changes.

29. Clasing, M., "Über den Einfluss von Oberflächenschichten, insbesondere Oxidschichten auf das Sintern von Metallen. II." *Z. Metallk.* 49:69-75 (1958).

The tensile strength, fracture strain, hardness, and density of sintered specimen, especially out of Cu, are investigated as a function of preoxidation and grain size of the powder, sintering temperature and sintering time. The different effects of sintering in vacuo and in H_2 are discussed.

30. Coble, R. L., and J. E. Burke, "Sintering in Crystalline Solids" in *Reactivity of Solids: Proc. 4th Intern. Symposium*, Elsevier Amsterdam (1960), pp. 38-51.

Forms useful review of recent sintering theory, and of relevant experimental work, as it concerns single-phase crystalline powder compacts, in which no liquid phase is formed and no pressure applied during heating. Three stages in sintering process discussed. Third stage involves the disappearance of isolated pores. Expression for rate of disappearance is given for cases in which the pores lie on grain boundaries; in other cases the rate is negligibly small, since grain boundaries provide the vacancy sink.

31. Cox, F. G., "Vacuum Sintering," *Metal Industry* (1960):186-189; 209-209.

32. Dan, G., and A. Protopopescu, "Electrical Resistivity as a Measure of Degree of Sintering," *Studii si Cercetari de Metalurgie* 4(4):353-364 (in Rumanian); *Rev. Roumaine de Metallurgie* 5(1):107-117 (1960) (in French).

This property can be measured with simple equipment. Electrolytic Fe powder, spectroscopically pure, of particle size below 60 μ was pressed without lubricant at 4 tons/cm². Two series of sintering experiments were carried out: (a) at temperatures from 500-1100°C; (b) the specimens were sintered successively, at temperatures rising in steps of 100° from 100 to 1100°C, with a sintering time of 1 hr at each temperature. Atmosphere of dried and purified H_2. Dimensional and weight measurements made in addition to electrical resistance. Special device for measuring very small resistances in question is described. Two sintering stages distinguished: (1) up to 500°C characterized by a rapid decrease of resistivity with rising temperature; (2) from 500-1100°C in which the decrease is very slow. Results are interpreted in terms of reduction of oxide films, metal–metal contacts, recrystallization, and increase of density.

33. DeHoff, R. T., *et al.,* "Size Sensitivity of Two-Particle Sintering Model," *Planseeber. Pulvermet.* 10(1/2):24-31 (Apr. 1962).

C. Herring's theory relating transport mechanism in sintering to exponent of scale factor in rate equation was re-examined by sintering contacting parallel copper wires in hydrogen at 930-1030°C; parameters measured were width of weld, wire diameter, and shrinkage of wire pair in direction of mutual diameters; it is concluded from results that exponent of scale factor is sensitive to wire diameter, and therefore does not identify transport mechanism in sintering.

34. DeHoff, R. T., *et al.*, "The Role of Interparticle Contacts in Sintering" in *Powder Metallurgy*, ed. by W. Leszynski, Interscience, New York (1961), pp. 31-51.

Topological model of sintering process leads to conclusion that rate of densification should be simply related to the genus, which is the difference between number of interparticle contacts and number of particles plus one. During first stage of sintering, it is found by experiment that the number of interparticle contacts may increase. Alteration of the specific genus by changing the number of particles in unit mass of the powder causes the rate of lineal contraction to vary inversely with the diameter. Alteration of the genus by rearranging the particle stacking causes the absolute rate of volume shrinkage to decrease linearly with increase in the average number of contacts per particle; at the same time, the fraction of the total possible shrinkage that is accomplished in unit time seems to be independent of the number of contacts per particle. These observations signify that the number of contacts per unit mass, rather than the neck geometry, is rate-controlling during early stage densification.

35. Dekeyser, W., "Lattice Defects and Reactivity of Solids" in *Reactivity of Solids: Proc. 4th Intern. Symposium*, Elsevier Amsterdam (1960), pp. 376-391.

Lattice defects may be defined as relatively permanent or transient departures, on the atomic scale, of the strict static order as required by lattice theory. According to this definition, the principal defects which can be present in a real crystal are point defects (vacancies, interstitials, atoms or ions of a different species), dislocations (edge or screw type), excitons, electrons and holes, thermal agitation, and finally the surface, or better, the superficial regions. A number of these may also be present in amorphous solids.

As many other properties, the reactivity of crystals depends on the presence of defects, and is therefore structure-sensitive. Reactions in which solids are involved may lead to the formation of a new phase outside on the surface of, or inside, the solid. In all cases a number of different types of defects are involved as well as interactions between them. The reactivity is not only determined by the superficial region, the defects in the bulk also play a role.

36. Dienst, W., and O. Werner, "Untersuchungen über die Grundlagen des Sintervorganges in Metallpulvern," *Z. Metallk.* 51:45-52 (1960).

The shrinking of Cu and Fe powder compacts in the direction of compacting in dependence of sintering temperature and time are graphically recorded. The graph, time vs. shrinking, may approximately be represented in specific time intervals by exponential functions, and the activation energy is determined. Accordingly, the most probable sintering mechanism of the early stage of sintering appears to be a plastic flow and for the later stages grain-boundary diffusion.

37. Dietze, H. D., "Zur Physik des Sintervorganges an Karbonyleisen," *Tech. Mitt. Krupp* 17:103-110 (1959).

Powder with spherical particles (mean size 2.7 μ) was pressed into cylinders under pressure of 2-8 M/cm^2 and the porosity p_p estimated from the density. Specimens were sintered in dry H$_2$ for 1-70 hr in the ranges 600-850°C and 950-1300°C, and the porosity p_s again determined. Micrographs of ground and polished specimens at 500× were enlarged 3× and the area and circumference of each pore were measured and the figures used to establish the mean curvature of a "rounding-off parameter" for each specimen. The decrease in porosity on sintering $(p_p - p_s) \propto p_p$.

38. Duffield, A., and P. J. Grootenhuis, "The Effect of Particle Size on the Sintering of Copper Powder," *J. Inst. Metals* 87:33-41 (1958/9).

Loose Cu powders of spherical form were sintered under vacuum at 650°C for 10 hr to form a porous material. Effects of varying particle size and size distribution within each batch of powder on the properties after sintering have been studied. The electrical conductivity was measured to determine degree of bonding throughout entire specimen. Measurement of tensile strength showed up the weakest section. The electrical conductivity was found to be independent of particle size distribution, but to vary with the mean size, even for specimens of equal densities. The tensile strength of porous material being obtained with powder of small size and very limited size distribution. A theoretical analysis of the flow of electricity through bonded spheres has been developed. The effect of variation in particle size can thus be predicted.

39. Eisenkolb, F., "Sintering in the Presence of a Liquid Phase" in *Powder Metallurgy*, ed. by W. Leszynski, Interscience, New York (1961), pp. 75-95.
      Object of investigation was behavior of metallic and nonmetallic additions which melt in the sintering process when mixed with metal powders. Sintering associated with formation of a liquid phase that consists of metallic additions allied to and alloying with the base metal is comparable to fusion welding, whereas the sintering by use of metals differing substantially from the base metal and alloying with it only on the surface is similar to soldering. Conditions such as bonding or cementing will result from the addition of nonmetallic substances, e.g., to the fluxing agent type or corresponding to the slag particles present in commercial metals. Starting with these considerations, tests were conducted, particularly on alloyed and unalloyed Fe powders varying in grain sizes and in origin from different production processes.

40. Elyard, C. A., "The Significance of Gas Adsorption Measurements in Relation to Residual Porosity," *Powder Metallurgy* (1963) No. 12:44-53.
      Measurement of the surface areas of UO_2 ceramics by gas-adsorption methods reveals marked differences between sintered and hot-pressed specimens. This difference can be accounted for by the existence of a very fine interconnected pore system in hot-pressed materials and of a relatively coarse interconnected pore system, together with completely closed pores, in sintered materials. Differences are also observed between the pore system in the center of a sintered material and that near its surface. Relative merits of various methods of measurement are discussed.

41. Eudier, M. A., "High-Density Sintering of Metal Powder Compacts" in *Powder Metallurgy*, ed. by W. Leszynski, Interscience, New York (1961), pp. 137-156.
      Brief review of different conventional ways of obtaining high-density parts: high temperature sintering, hot pressing, hot forging, rolling, extruding, and drawing. Comparisons made to explain different practical results obtained, and ways of improving them. Other possibilities discussed, especially use of ultrafine powders, special atmospheres on thin film, and intermediate liquid phases created by small additions such as phosphorus which can be eliminated afterward. One of the examples is related to the Bi_2Te_3 semiconductors.

42. Fedorchenko, I. M., and R. A. Andrievskii, "On the Mechanism of Sintering in Single-Component Systems," *Poroshkovaya Metallurgiya* 1(1):9-18 (Jan.-Feb.1961). 1961).
      Corresponds in form and content to a major part of paper in *Powder Metallurgy* (1959) No. 3:147-171. Report given of author's work on flow processes in sintering, and effect of surface condition of particles. Densification in sintering held to take place by plastic flow (by one school of thought) and by another to come about by diffusion. Occurrence of plastic flow in sintering not proved by direct experiment.

43. Fedorchenko, I. M., and N. V. Kostyrko, "The Mechanism of Shrinkage on Metal Powder Compacts during Sintering," *Fiz. Metallov i Metallovedenie* 10:75-83 (1960);

English translation: *Physics of Metals and Metallography* published by Acta Metallurgica.

Effect of annealing on the specific surface area of metal powders and relationship between it and the degree of shrinkage during sintering of compacts studied on samples both of pure metals and mixtures. In every case specific surface area decreased after preliminary annealing, which also brought about a considerable decrease in degree of shrinkage during subsequent sintering. The fact that the latter effect was much more pronounced in case of powder mixtures is attributed to heterodiffusion taking place in surface layers of the powder particles during annealing. A very slight increase in the pycnometric density of the powder particles after annealing at various temperature indicated that shrinkage of sintered powder compacts cannot be attributed to increase in density of the individual particles due to decrease in the degree of internal porosity and the concentration of structural defects. Consequently, the predominant part in sintering shrinkage is played not by volume diffusion, but by those different processes which bring about reconstruction of the surface layers of the powder particles. By allowing these processes to take place during preliminary annealing, the risk of excessive dimensional changes of powder compacts during sintering can be greatly minimized.

44. Fischmeister, H., and R. Zahn, "Model Experiments on the Sintering Mechanism in High-Purity Iron," *Ber. über die II. Internationale Pulvermet. Tagung in Eisenach* Akad. Verlag, Berlin (June 1961), pp. 93-98.

Authors show why with sintering model used by Kuczynski and general exponential rate equation, the mathematical analysis of sintering kinetics can only be very approximate, so that values obtained from experimental results for the diffusion constant for any particle transport mechanism is very much open to question. They propose to limit application of results of model experiments to comparison of the Kuczynski exponent n, and the activation energy with the values to be expected from theoretical considerations.

45. Fraunberger, F., and A. Külb, "Study of Sintering Processes in Mixtures with a Magnetic Component by the Alternating Field Method," *Z. Metallk.* 50(3):179-181 (1959).

To determine concentration distribution maxima and sintering time required to reach a given state, authors recommend method in which the much more sharply defined "Hopkinson temperatures," in place of Curie temperatures, which also stand in a linear relationship to the concentration, are measured. Basis of method explained, and practical application described.

46. Fraunberger, F., and V. Scheuing, "Course of Concentration Changes during Sintering of Nickel-Iron and Nickel-Copper Powder Mixtures," *Z. Metallk.* 52 (8):547-550 (1961).

"Alternating field method" applied to following the course of diffusion during sintering of Fe-Ni powder mixtures. Results presented for mixtures of different compositions, showing effects of sintering time, temperature, and compacting pressure. Shortcomings and advantages of "alternating field method" discussed.

47. Geach, G. A., "Recent British Developments in the Theory of Sintering," *Powder Metallurgy* (1959) No. 3:104-114.

Review work related to theory of sintering carried out in United Kingdom since that published at Symposium on Powder Metallurgy (1954). British Standards Institution definition of "sintering"; discussion of sintering process; nature of material and sintering; active materials; stoichiometry; sintering atmosphere; grain boundaries.

48. Geach, G. A., and A. A. Woolf, "The Sintering Behavior of Organic Materials" in *Powder Metallurgy*, ed. by W. Leszynski, Interscience, New York (1961), pp. 201-206.

Number of mechanisms, including creep, required to explain sintering of materials. Purpose of this work was to examine whether sintering and creep processes are necessarily related. A range of organic molecules and one organometallic compound, whose sintering temperature would be expected to be greater than room temperature were examined. Qualitative tests made on powders and quantitative tests on pressed rods, but in no instance was true sintering observed even at temperatures above 75% of their absolute melting point. However, the two substances which were examined underwent creep at these temperatures.

49. Geguzin, Ya. E., "Study of the Sintering of Metal Powders at a Constant Heating Rate," *Fiz. Metallov i Metallovedenie* 6(4):650-656 (1958); English translation: *Physics of Metals and Metallography* published by Acta Metallurgica.

Previous "isothermal" sintering shrinkage curves are in reality not so, since an appreciable part of the shrinkage has taken place during the heating-up of the specimen to the temperature in question. Pines pointed out the connection between kinetics of shrinkage and removal of distortions of the crystal lattice of the powder particles. The removal of lattice distortions on rise of temperature will depend on rate at which temperature rises. Experiments made on "active" Cu powder. Curves show variation of shrinkage with duration of heating and the temperature. Coefficient of self-diffusion as a function of temperature passes through maximum which is displaced toward the region of higher temperature.

50. Goodison, J., and J. White, "The Sintering of Mixed Powders" in *Agglomeration*, ed. by W. A. Knepper, Interscience, New York (1962), pp. 251-268.

Sintering behavior of mixed oxide powders investigated with reference to nature of mechanisms that affect densification. Where two components form solid solutions, particles of the slow-diffusing component tend to increase in size. When components form a compound, densification is frequently inhibited. Where the components are not appreciably soluble in each other, densification may be inhibited presumably because of high interfacial tension between particles.

51. Guiraldenq, P., et al., "Comparison of the Phenomena of Autodiffusion in Volume and Intergranular Diffusion in Sintered Iron as Well as in the Solid Material," *Compt. rend.* 254:99-101 (Jan. 3, 1962).

Investigation of the self-diffusion of alpha and gamma iron made by conventional and powder metallurgy methods, respectively, both of identical impurity content. Hardly any difference exists in the alpha range volume diffusion. In the gamma range the volume diffusion is considerably higher for the sintered material. Intergranular diffusion was considerably higher for the sintered iron in the alpha as well as in the gamma range.

52. Hausner, H. H., "Sintering Behavior of Loose Metal Powders," *Planseeber. Pulvermet.* 9(1/2):26-32 (Apr. 1961).

In pressureless compacting, the density of a sintered mass of loose powders depends considerably more on particle size, particle size distribution, and particle activity, than do pressure-compacted powders. Examples are given for spherical and irregularly shaped stainless steel powders. The powder composition which results in the greatest apparent density does not necessarily sinter to the highest sintered density. It has been shown that rate of sintering depends on particle size distribution as well as on the activity of the particles.

53. Hausner, H. H., "The Effect of Porosity of Compacts on the Structure of Sintered Metals," *J. Japan Soc. Powder Met.* 7(2):58-72 (Apr. 1960); *Trans. Indian Inst. Metals* 13:351-373 (Dec. 1960) (English).

Author lists and discusses eleven variables characterizing porosity in a metal powder mass, which may be loose or compacted, unsintered or sintered. Analysis of effects of some of these variables on rate of densification enables derivation of equation expressing that rate entirely in terms of porosity. Changes in porosity during sintering can be interpreted in terms of the migration of lattice vacancies. Comparison of loose and pressure-compacted powders with regard to sintering behavior.

54. Hausner, H. H., "Definition of Sintering in Powder Metallurgy," *Planseeber. Pulvermet.* 11(2):59-69 (Aug. 1963).

Analysis of various definitions. Analysis of sintering process. Proposition for a new definition and discussion of it by scientists from various countries.

55. Hausner, H. H., "Basic Studies in Linear Shrinkage Behavior during Sintering" in *Progress in Powder Metallurgy-1963, Vol. 19,* Metal Powder Industries Federation, New York, pp. 67-85.

One of the fundamental problems in powder metallurgy is densification or shrinkage of mass of metal powder during sintering. Paper shows effects of particle size, compacting pressure, and sintering conditions on linear shrinkage behavior in the direction of compacting and perpendicular to this direction. It is shown that shrinkage behavior is different during sintering at low temperature than at high temperature, that fine powder particles shrink differently in various directions than coarse powder compacts, and that the shrinkage behavior depends frequently on the powder particle size distribution. Further, with linear shrinkage of compacts sintered under tensional load as well as of compacts which are exposed to thermal cycles during sintering.

56. Hausner, H. H., "Compacting and Sintering of Metal Powders without the Application of Pressure" in *Agglomeration,* ed. by W. A. Knepper, Interscience, New York (1962), pp. 55-88.

Deals with basic differences between masses of loose or pressureless compacted metal powders and those which are compacted under pressure, such as deformation and nondeformation of particles, presence and absence of stresses in particles, and additional stresses due to pressure compacting; differences between type of contact between the particles, orientation of particles and voids, differences in the type of porosity. Emphasis given to particle size distribution and various types of distribution curves, effect of which is more important in pressureless than in pressure compacting.

57. Hayden, H. W., and J. H. Brophy, "Activated Sintering of Tungsten with Group VIII Elements," *J. Electrochem. Soc.* 110(7):805-810 (July 1963).

Small additions of Group VIII transition elements permit densification of W powder compacts at temperature well below those employed for sintering commercial purity W powder; of added elements explored to date, Pd appears to have greatest effect, followed in order by Ni, Rh, Pt, and Ru.

58. Hornstra, J., "The Role of Grain Boundary Motion in the Last Stage of Sintering," *Physica* 27:342-350 (1961).

The last stage of sintering is governed by diffusion of vacancies from pores to grain boundaries. Precipitation of vacancies at boundaries causes elastic stresses, which make the rate of precipitation uniform over the boundary area. This is not true when grains are sliding along the boundary and the boundary is suitably curved. Then most vacancies can precipitate in the neighborhood of the pores and rate of sintering will be higher than in the case of a plane boundary. This shows that plastic deformation can play a role in sintering, although vacancy diffusion remains the most important process. The dislocation model of grain boundaries is used to illustrate the role of vacancy diffusion in grain boundary motion. The dependence of the overall rate of sintering on pore size distribution is given in appendix.

59. Ichinose, H., and G. C. Kuczynski, "Role of Grain Boundaries in Sintering," *Acta Met.* 10:209-213 (Mar. 1962).

Effect of presence of grain boundaries on sintering rate investigated by measuring growth of necks between three twisted wires as function of time and temperature. It was found that, in presence as well as absence of grain boundary in the neck, volume diffusion controls the process. The measured rate of neck growth was in good agreement with that calculated from theory. Rate of approach of centers of wires also studied. In absence of grain boundaries, no approach of centers was observed. In presence of grain boundaries the centers did approach at a rate which was measured. Study allowed distinction between grain boundary and surface sinks of vacancies during sintering; in absence of grain boundaries, diffusion flux of vacancies is from region just beneath neck to adjacent free surface; in presence of grain boundaries, vacancies migrate from the region just beneath neck to grain boundary.

60. Ivensen, V. A., "On the Diffusion Theory of Sintering," *Fiz. Metallov i Metallovedenie* 6(2):370-375 (1958); English translation: *Physics of Metal and Metallography* published by Acta Metallurgica.

According to diffusion theory of Pines *et al,* elimination of pores during sintering takes place by formation of vacancies on internal concave surfaces bounding the pores, and the diffusion of these vacancies to external surface of compact. This theory disputed by Ivensen. Mackenzie and Shuttleworth pointed out that densification by means of diffusion process would be extremely slow. High rate of densification at beginning of sintering is explained according to diffusion theory of Pines, by presence of lattice distortions existing in real crystals, as distinct from ideal crystals from which the slowness of diffusion was deduced. In such real crystals, process of recovery during sintering is held to result in formation of additional vacancies; thus to increase rate of diffusion. This is found by Ivensen to contradict one of basic ideas of diffusion theory, according to which rate of flow of material depends not on total number of vacancies, but on concentration gradient. Further objection is that a mechanism which determines rapid rate of densification in early stages characterized by communicating or "open" porosity would not cease to operate when the porosity becomes closed. Frenkel's theory is held to account more satisfactorily for the phenomena of sintering in real particles, although it does not reveal the atomic mechanism of the process, and thus has a "semiphenomenological" character.

61. Iwase, K., *et al,* "Sintering and Crystallographic Characteristics of Very Fine Oxide Particles Produced by Calcination of Salts" in *Powder Metallurgy,* ed. by W. Leszynski, Interscience, New York (1961), pp. 173-189.

By electron microscopy and electron diffraction, as well as X-ray analysis, sintering of various kinds of oxide and metallic particles and their crystallographic structures were studied. All oxide particles prepared by calcination of metallic salts. Metallic particles obtained by reduction of same salts. Results of electron microscopic observations show that ultra fine oxide particles appear as first precipitated in mother salts during calcination and also are linked with each other at points or areas of contact, forming aggregates whose shape resembles that of the mother crystal. In electron diffraction photographs of most of these aggregates, there appear net-pattern configurations which are so clear that unit particles in the same aggregate can be considered to be highly oriented. This orientation seems to be greatly influenced by pre-existing crystallographic structure of mother salts and plays a very important role in early stage of subsequent sintering. Shape and structure of the aggregate are rather well maintained. On further sintering, however, the above mentioned shape and structure of aggregate inherited from the mother crystals change gradually.

62. Jackson, J. S., "Hot Pressing High-Temperature Compounds," *Powder Metallurgy* (1961) No. 8:73-100.

Technique of hot pressing and factors to consider in designing suitable equipment are discussed; examples quoted of three main classes of materials: unbonded hard materials, materials containing small amount of cementing phase or impurity affecting densification, and materials containing appreciable amount of cementing phase; current theories concerning mechanism of densification during hot pressing are considered.

63. Kingery, W. D., "Sintering in the Presence of a Liquid Phase" in *Ceramic Fabrication Processes,* John Wiley, New York (1958), pp. 131-143; see also in *Kinetics of High-Temperature Processes,* John Wiley, New York (1959), pp. 187-194.

Considers surface tension forces and attempts to explain their functions (using pure liquid sintering as example). Attempt to estimate amounts of liquid phase which may be present in firing ceramic ware; finally considers case of nonreactive liquid–solid system and a reactive liquid–solid system.

64. Kothari, N. C., "Sintering Kinetics in Tungsten Powder," *J. Less Common Metals* 5(2):140-150 (Apr. 1963).

Experimental study of initial sintering kinetics in W powder made at 1100-1500°C; influential factors involved include compacting pressure, sintering time and temperature; isothermal changes in volume and density were studied; experimental results evaluated by two methods of analysis; ratios between activation energy of surface diffusion and volume self-diffusion and between grain boundary diffusion and volume self-diffusion were examined in W and other metals; they support theory that sintering of W powder at 1100-1500°C is controlled by grain diffusion and not volume self-diffusion.

65. Kovalchenko, M. S., and G. V. Samsonov, "Application of the Theory of Viscous Flow to the Hot Pressing of Powders," *Poroshkovaya Metallurgiya* 1(2):2-13 (Mar.-Apr. 1961).

Consolidation of powders by hot pressing can be described in terms of bulk viscous flow of a porous body under the action of surface tension and external pressure. The process is analyzed from this point of view, and expressions for rate of sintering are obtained by equating the work of external forces with that of the forces of internal friction. It is shown that in the case of crystalline substances there is a linear relationship between the time of sintering and the viscosity of the compact material. The theoretical relationships derived were checked with the aid of experimental data obtained from hot pressing of WC powder. Values determined for viscosity of material at sintering temperature and energy of "loosening" of the WC lattice.

66. Kuczynski, G. C., "Effect of Oxygen on Sintering of Oxides" in *Powder Metallurgy in the Nuclear Age,* Metallwerk Plansee, Reutte, Austria (1962), pp. 166-179.

Recent experiments on sintering of Al_2O_3, Fe_2O_3, and ZnO revealed that oxygen present in the ambient atmosphere may increase significantly the apparent diffusion coefficients in oxides and thus increase appreciably sintering rates. This is probably due to increase of the point defect concentration in surface layer, due to the oxygen adsorption. This effect increases the diffusion coefficients in the presence of defect gradients existing in the necks between the particles or around the small pores.

67. Kuczynski, G. C., "Theory of Solid State Sintering" in *Powder Metallurgy,* ed. by W. Leszynski, Interscience, New York (1961), pp. 11-29.

Present status of theory reviewed. As details of the geometry of the contacts between individual particles in powder compacts are too complex to be treated theoretically, discussion limited to simplified models such as spheres to spheres and spheres (or wires) to plates. Experiments with such samples yielded fairly reliable confirmation of theory proposed by the author. According to this theory, the first stage of sintering, characterized by formation of a neck between two particles, can be brought about by one or more of the following processes: viscous or plastic flow, evaporation and condensation, and volume diffusion or surface diffusion. General equation given for well-known relationships between the radii x of the neck and a of the spheres at time t, and temperature T. Process of densification, or closure of isolated pores, less extensively investigated. Evidence so far amassed indicates this process is closely related to Nabarro–Herring microcreep.

68. Kuczynski, G. C., "Theory of Residual Porosity in Powder Compacts," *Powder Metallurgy* (1963) No. 12:1-16.

Problem of residual porosity in relation to grain growth in powder compacts is discussed. It is shown that present diffusion theory of densification, together with Zener's relation, can account, at least qualitatively, for the residual porosity in powder compacts. Moreover, this theory brings out the physical meaning of the so-called "Sauerwald temperature" phenomenon. The porosity occurring in multicomponent powder compacts, as a result of the difference in diffusivities, is also discussed.

69. Kuczynski, G. C., and P. F. Stablein, Jr., "Sintering in Multicomponent Systems" in *Reactivity of Solids: Proc. 4th Intern. Symposium,* Elsevier Amsterdam (1960), pp. 91-104.

Phenomena accompanying sintering in a Cu–Ni system investigated on synthetic wire compacts. It was found that interdiffusion predominates during first stage of process. It arrests neck growth until the chemical gradient is leveled out. During this stage compact actually expands. This is chiefly due to Kirkendall–Hartley effect, which introduces large volume changes in compact. Precipitation of vacancies as well as plastic flow caused by this effect have been observed. At lower temperature the surface diffusion appears to be predominant.

70. Lenel, F. V., "Fundamentals of Powder Metallurgy—Creep in Powder Metallurgy Products" in *Proc. 17th Annual Meeting Metal Powder Industries Federation* (Apr. 1961), pp. 108-119.

Theory of creep and its applications to steady-state creep rate of certain powder metallurgy products; slip at low temperature, types of creep; creep in dispersion-strengthened materials governed by rates of recovery; creep in recrystallized SAP, in dispersion of strengthened lead, in fine grained SAP, and in zinc powder and in zinc alloy powder extrusions.

71. Lenel, F. V., *et al,* "Influence of Gravity in Sintering," *Trans. Met. Soc. AIME* 227(3):640-644 (June 1963).

Radial shrinkage during sintering of cylindrical compacts and loose aggregates of Cu powder was found to be nonuniform from top to bottom of samples and to depend on method of supporting them; nonuniformity is ascribed to effect of gravity forces; it is concluded that, since gravity has effect in sintering without externally applied stresses, no sharp dividing line can be drawn between conventional sintering and hot pressing.

72. Lenel, F. V., *et al,* "Driving Force for Shrinkage in Copper Powder Compacts during Early Stages of Sintering," *Powder Metallurgy* (1962) No. 10:190-198.

Shrinkage behavior of compacts from irregular copper powder during initial stages of sintering determined by dilatometric method; effects of compacting pressure and of external load during sintering at constant heating rate of 3°C/min upon shrinkage were observed; same rate to temperatures of 200, 300, 400, 500, and 600°C also measured.

73. Lenel, F. V., *et al,* "Some Observations on Shrinkage Behavior of Copper Compacts and of Loose Powder Aggregates," *Powder Metallurgy* (1961) No. 8:25-36.

Experiments led to conclusion that action of surface tension forces cannot adequately explain observed ratios of radial and axial shrinkage either in compacts or in loose powder aggregates; it is tentatively postulated that gravity is responsible for observed ratio of shrinkage in loose powder aggregates, while residual stresses are responsible for shrinkage effects in compacts with interconnecting pores.

74. Lenz, W. H., and J. M. Taub, "Liquid Oxide Phase Sintering of Molybdenum and Tungsten," *J. Less Common Metals* 3(5):429-432 (Oct. 1961).

In hydrogen sintering of Mo and W at about 1700°C, compacts containing minor amounts of TiO_2 combined with UO_2 or ThO_2 will permit sintered density to approach theoretical; combination of these oxides appears to result in phase which is liquid at sintering temperature, and wets Mo and W.

75. Lenz, W. H., and P. R. Mundinger, "High-Density W–UO_2 by Activated Sintering," *J. Less Common Metals* 5(2):101-111 (Apr. 1963).

Reference made to increasing interest in combining nuclear fuel materials with refractory metals for high temperature applications; activated sintering of Ni-doped W powders in hydrogen was confirmed at as low as 1100°C; presence of considerable UO_2 caused Ni additive to lose much of its effect; relative merits of Ni and/or TiO_2 additives as activators in W–50 v/o UO_2 mixtures investigated with respect to particle size and sintering cycle; at 1700°C some of these compositions could be sintered to near theoretical density in 1-4 hr; effect of H_2 and vacuum atmospheres on sintered density of these materials compared with similar effects observed for W and Re powders.

76. Likhtman, V. I., and M. L. Smolyanskii, "Physicochemical Phenomena in the Pressing and Sintering of Metal Powders," *Uchenye Zapiski, Moskov. Gosudarst. Pedagog. Inst. im. M. N. Pokrovskogo* 1962, No. 9:46-62. *Referat. Zhur. Met. (Moscow)* 1963 (3):G.297.

Stresses importance of studying change in contact surface area during pressing and sintering. Explains briefly action of surface-active lubricants in plastic deformation of metals. In sintering of Cu powder compacts, change of contact surface area (investigated by way of electrical conductivity) passed through three stages: (1) relaxation of residual stresses, resulting in rupture of some contacts; (2) reduction of oxides; (3) selective recrystallization in and between particles.

77. Lund, J. A., *et al,* "Studies of Sintering and Homogenization of Nickel–Copper Compacts," *Powder Metallurgy* (1962) No. 10:218-235.

Studies of progress of sintering and alloying in compacts of similar compositions made from Ni-coated Cu, Cu-coated Ni, and mixed Ni and Cu powders; alloying by diffusion at both 1900 and 2200°F progressed most rapidly in compacts prepared from Ni-coated Cu powders; electrical resistivity used to follow homogenization of compacts, and samples rendered nearly 100% dense by cold working and annealing before making resistivity measurements; observations discussed in detail.

78. McClelland, J. D., "Kinetics of Hot Pressing" in *Powder Metallurgy,* ed. by W. Leszynski, Interscience, New York (1961), pp. 157-170.

Hot pressing of powder materials can be adequately described using a plastic flow theory proposed by Mackenzie and Shuttleworth for sintering. Present derivation assumes that principal driving force for closing of pores is the applied hydrostatic pressure instead of the surface tension of the pores. This pressure term, when corrected for the density of the compact, can be substituted for the surface tension term in the original derivation. Resultant equations give rate of densification of material in terms of a yield point and a viscosity. The equations predict that an endpoint density will be reached which is dependent on applied pressure and the yield point of the material.

79. McIntyre, R. D., "The Effect of HCl–H₂ Sintering Atmospheres on Properties of Compacted Tungsten Powder," *Trans. Quart. ASM* 56(3):468-476 (Sept. 1963).

Investigated for temperatures from 1400-1800°C and sintering times from 1/2 to 8 hr. Gas composition varied from dry H₂ to 30% HCl in H₂. Presence of HCl in H₂ atmosphere increased densification, promoted void spheroidization, caused a more uniform void distribution and lowered residual oxygen in the powder. Both tensile strength and elongation were improved for sintered W compacts at equivalent sintering times and temperatures. Improvement by sintering by addition of HCl to the dry H₂ furnace atmosphere was probably result of more active surfaces which result. Approximate activation energies for H₂ and HCl–H₂ sintering were 72,000 cal/mole and 52,000 cal/mole, respectively.

80. McQueen, H. J., and G. C. Kuczynski, "Sintering of Zinc Sulfide," *J. Am. Ceram. Soc.* 45(7):343-346 (July 1962).

Mechanism of sintering ZnS studied by measuring rate of growth of the necks between polycrystalline spherical particles prepared by grinding sintered ZnS in a small air-operated grinder. For sintering in vacuum, mechanism at temperatures above 600°C was found to be volume diffusion, while below that temperature, it was surface diffusion. In an atmosphere of Zn vapor, surface diffusion mechanism was found to predominate up to 650°C. In Ar, at 750°C, sintering occurred by volume diffusion.

81. Masuda, Y., "Calorimetric Study on Sintering of Copper Powder Compact," *Tohoku Univ. Research Inst. Science Report Ser. A* 14(3):156-164 (June 1962).

Measurement of change in specific heat of compact during sintering, using Nagasaki–Takagi procedure; process accompanied by energy release between 230-520°C and energy absorption above 650°C; thermal energy released was 2.46 cal/g, which corresponds to annihilation of vacancies of $7 \cdot 10^{-3}$/atom. It is concluded that predominant mechanism of material transfer would be surface diffusion or like mechanism associated with vacancy migration.

82. Margerand, R., and M. Eudier, "Elimination of Porosity by Sintering," *Powder Metallurgy* (1963) No. 12:17-26.

The behavior of pores in metal powder compacts during solid-phase sintering is examined, and theoretical considerations compared with practical observations. Elimination of porosity is possible in very thin specimens by selecting a relatively fine powder. With thicker compacts it is also necessary to bring the specimen very slowly up to sintering temperature. When the dimensions exceed a few centimeters, however, porosity can only be eliminated by increasing the rate of sintering by adding another metal; conditions governing such additions are defined.

83. Martin, A. J., and G. C. Ellis, "The Relationship between Powder Properties and Sintering Behavior of Beryllium," *Powder Metallurgy* (1961) No. 7:120-138.

Surveys existing routes available for preparation of beryllium powders and comments on effect of certain properties, such as particle size distribution, on compaction of powders by different techniques. It is emphasized that particle size, degree of oxidation, and overall purity are interdependent variables and that little attempt has so far been made to study their effects separately.

84. Matsumura, G., "Swelling of an Iron Copper Compact during Sintering," *Planseeber. Pulvermet.* 9 (1/2):33-35 (Apr. 1961) (English).

Describes model experiments using Fe wires of 1.2 mm diameter, which were dipped in vacuum into molten Cu saturated with Fe, and immediately cooled to room temperature. When the temperature was again raised, there was diffusion of the Cu into the Fe, but no diffusion of Fe into Cu, this being already saturated with Fe. The wires were heated at 1050°C in H_2 for periods up to 200 hr. The increase in diameter of the wires with heating time was measured microscopically. Swelling continued for 50 hr, after which the diameter of the wires remained constant. Even after heating for 200 hr, the Fe was not completely saturated with Cu. As a result of the one-way diffusion, the Cu–Fe interface moved toward the Cu side, and a gap was formed, by the Kirkendall effect, between the two. In an Fe–Cu powder compact, with its different structure, "it is assumed" that each Fe particle continues to swell, without cracking, until it is saturated with Cu.

85. Mitsche, R., *et al.*, "Direct Observation of the Sintering Process by the High-Temperature Microscope" in *Powder Metallurgy in the Nuclear Age,* Metallwerk Plansee, Reutte, Austria (1962), pp. 799-829 (in German).

With the new high-temperature equipment for metal microscopes (C. Reichert, Vienna), sintering reactions in various systems preferably such involving metals, at temperatures up to 1800-2000°C, were visually followed and also photographically recorded. The results with photos and motion pictures are discussed.

86. Morgan, P. E. D., and A. J. E. Welch, "The Sintering of Pure and 'Doped' Metal Oxides" in *Reactivity of Solids: Proc. 4th Intern. Symposium,* Elsevier, Amsterdam (1960), pp. 105-111.

Isothermal shrinkage of metal–oxide compacts at temperatures up to 1400°C has been measured by means of a dilatometer, with which linear dimensional changes could be continuously observed. Pure oxides studied, selected to bring out the broad effects of crystal structure and to avoid undue complexities due to nonstoichiometry, were MgO, Al_2O_3, SnO_2, TiO_2, and Nb_2O_5. MgO, Al_2O_3, TiO_2, and Nb_2O_5 show appreciable crystal growth only after shrinkage to high bulk densities; SnO_2, possibly on account of its volatility, crystallizes at low bulk densities. Crystallization and other textural changes were observed by X-ray measurements of crystal size, and by microscopical examination.

87. Murray, P., "The Properties of Active Ceramic Oxide Powders in Relation to Sintering Behavior" in *Agglomeration,* ed. by W. A. Knepper, Interscience, New York (1962), pp. 93-111.

Properties of chemically prepared oxide powders described with particular reference to surface area, crystallite size, microstructure, and packing properties. Experimental data relating these properties to subsequent sintering behavior are given with an appraisal of state of knowledge in this field.

88. Nicholas, M. G., "Cold Bonding between Hemispherical Copper Surfaces," *Trans. Met. Soc. AIME* 227 (1):250-256 (Feb. 1963).

Reviews work on interparticle bonding in unsintered powder compacts. Evaluation of influence of individual factors and quantitative correlation with overall strength of compact are impossible, owing to complex geometry of the compact and the simultaneous occurrence of different events. Author's experiments using a macroscopic model consisting of a pair of hemispherically-ended Cu rods of 0.25-in. diameter are described. Factors studied were influence of deformation alone, of deformation plus relative movement, and of surface contamination. It was found that no bonding occurred when clean surfaces were merely pressed together, even with contact pressures of 50,000 psi. Bonding resulted when clean surfaces were simultaneously pressed and moved relative to each other, no matter how small the pressure or relative movement.

89. Oel, H. J., "*Zur Thermodynamik und Kinetik des Sinterns,*" *Z. Metallk.* 51:53-58 (1960).

Experiments with model bodies and with powder for clarification of the sinter mechanism of single material systems are reported. Various measuring methods for determining reactivity of powder with respect to sintering are compared.

90. Oel, H. J., "Das Sintern und die Eigenschaften von Pulvern," *Ber. über II. Intern. Pulvermet. Tagung in Eisenach* (June 1961), pp. 85-92.

Describes X-ray methods for determining activity of grain boundaries in powder particles. Tests made with MgO powder, and concern mainly the vacancy concentration in the grain boundaries.

91. Oel, H. J., "The Relationship between Free Energy and Kinetics in Sintering Processes" in *Kinetics of High-Temperature Processes,* John Wiley, New York (1959), pp. 179-186.

Both the free energy of the substance and the kinetics of sintering process depend on number of defects. Experiments show the different behavior of amorphous and regular powders.

92. Oel, H. J., "Characterization and Sintering of Powders" in *Agglomeration,* ed. by W. A. Knepper, Interscience, New York (1962), pp. 271-298.

Empirical experience in working of ceramic and metallurgical powders, experimental investigations, and theoretical considerations show that the particle size, as defined and determined by sedimentation or other methods of measurement, is not sufficient to characterize a powder. Especially in regard to the pressing and sintering behavior of powders, a consideration of their point line and two-dimensional defects (vacancies, dislocations, and crystallite boundaries) is highly important because the processes of pressing and sintering are only made possible by their presence; they supply the means for transportation of the material.

93. Okamoto, Y., "Study of Phenomena in Liquid-Phase Sintering," *J. Japan Soc. Powder Met.* 9(1):1-6 (Feb. 1962); 1-9 (Apr. 1962).

Changes of microstructure, particle growth, and rate of densification during sintering of Cu–Ag in the presence of liquid phase were observed; rapid densification occurred in early stage of sintering, and increased with increase of liquid phase and temperature; particle growth by solution precipitation process was noted. Effects of liquid solubility and wettability on microstructures and densification during sintering of Cu–Pb, Cu–Bi, Fe–Cu, Fe–Ag, Co–Cu, and Co–Ag systems in presence of liquid phase are considered.

94. Parravano, G., "Chemical Sintering of Berthollide Compounds" in *Reactivity of Solids: Proc. 4th Intern. Symposium,* Elsevier, Amsterdam (1960), pp. 83-90.

Role of solid state reactions in sintering of berthollide compounds is examined. Under some conditions, free energy change of chemical transformation is the determining factor for welding together of particulate nonstoichiometric solids. A diffusional transfer of matter in the solid controls the chemical reaction and, consequently, sintering. Mathematical expressions for diffusion and sintering in the simple case of two spherical particles are presented and results compared with experimental data on sintering of zinc oxide microspheres. Equations for shrinkage of pores as a result of a chemical reaction are also presented.

95. Pines, B. Ya., "Further Discussion of Theories of Sintering," *Fiz. Metallov i Metallovedenie* 6(2):375-381 (1958); English translation: *Physics of Metals and Metallography* published by Acta Metallurgica.

Author replies to criticism of V. A. Ivensen and claims that diffusion theory should not be set against the viscous flow theory of Frenkel, but is a development of it, so that the antithesis found by Ivensen is invalid. Although this "diffusional variant" of the Frenkel theory must not be regarded as perfected, and requires further elaboration, its agreement with facts is held to leave no doubt regarding the correctness of its basic positions.

96. Pines, B. Ya., and A. F. Sirenko, "Study of Diffusion Creep in Sintering of Metal Powders," *Issledovaniya po Zharoproch. Splavam* 4:301-310 (1959).

Sintering of metal powders offers favorable opportunity for study of "creep," because we can compare two manifestations of it, due respectively to external forces and to surface tension at the surfaces of the internal pores. It is also of interest to elucidate the nature of creep in nonequilibrium materials and nonhomogeneous bodies. In different temperature ranges and under different stress conditions, it is possible that different mechanisms of creep operate. In particular, there is apparently "dislocation" creep and "diffusion" creep. Sintering may be regarded as a manifestation of diffusion creep under the action of surface tension forces. In order to approximate conditions of diffusion creep in which deformation is due to flow by means of a directed self-diffusion of atoms, the tests were carried out at relatively high temperatures and under small applied loads. It is found that at high temperatures the course of deformation due to creep in sintered bodies does not differ from that in solid bodies.

97. Pines, B. Ya., and A. F. Sirenko, "Nonequilibrium Conditions and Diffusion Creep in Sintered Compacts," *Fiz. Metallov i Metallovedenie* 7(5):766-776(1959); English translation: *Physics of Metals and Metallography* published by Acta Metallurgica.

Previous work led to conclusion that sintering takes place by way of self-diffusion, the migration of atoms from the outer surface resulting in closing of pores. The process may also be looked upon as a manifestation of diffusion creep under the influence of surface tension forces. Sintering rate, however, is found to be two to three times that predicted on basis of diffusion creep. It is shown that this must indicate an increased value of the coefficient of self-diffusion, which, according to theory of Frenkel, may be due to the presence of excess vacancies in the crystal lattice, caused by the gradual elimination of lattice defects and distortions at elevated temperatures.

98. Pranatis, A. L., and L. Seigle, "Sintering of Wire Compacts" in *Powder Metallurgy,* ed. by W. Leszynski, Interscience, New York (1961), pp. 53-71.

The influence of the presence of grain boundaries on rates of sintering of Ni, Cu, and Fe wire-wound compacts is studied. In Fe compacts sintered in the alpha range, surface diffusion appears predominant and no grain boundary effect is therefore expected or evident. On other hand, sintering of Cu and Ni compacts at temperatures close to their respective melting points, apparently occurs by volume diffusion, but with varying grain boundary effects. Neck growth in both materials appears entirely unaffected by presence or absence of grain boundaries. Void shrinkage was found to be sharply inhibited in Ni samples from which grain boundaries had been deliberately eliminated, but no definite effect was noticed in Cu, in apparent contradiction to previous results.

99. Pranatis, A. L., *et al, Fundamentals of Sintering. III,* U. S. AEC Report SEP-250 (June 1958).

Influence of presence of grain boundaries on rates of sintering of Ni, Cu, and Fe wire-wound compacts. In Fe compacts sintered in alpha range, surface diffusion appears predominant and no grain boundary effect is evident. Sintering of Cu and Ni compacts at temperatures close to their respective melting points apparently occurs by volume diffusion, but with varying grain boundary effects.

100. Pranatis, A. L., *et al.*, *Fundamentals of Sintering. II*, U. S. AEC Report SEP-247 (Feb. 1958).

Rates of spheroidization and densification of Au and Ni wire compacts were measured as function of time, temperature, and wire size. The necks joining the sintered particles were found to grow proportionally to the fifth root of time. The value of the coefficient of self-diffusion calculated from the sintering rate is of the same order of magnitude as that obtained by ordinary tracer techniques but with a somewhat higher activation energy. These results are entirely consistent with those predicted by the Kuczynski diffusional theory of sintering, as is the effect of wire size on sintering rate.

101. Pranatis, A. L., *et al.*, *Fundamentals of Sintering. I*, U. S. AEC Report SEP-229 (Oct. 1956).

102. Raichenko, A. I., and V. V. Skorokhod, "Theory of Shrinkage in the Initial Stages of Sintering," *Poroshkovaya Metallurgiya* 1(4):3-8 (July-Aug. 1961).

Authors aim to develop physical theory of shrinkage based on theory of crystal lattice defects, taking into account various kinetic processes which take place in real porous bodies. Shrinkage of such bodies is considered to occur by "solution" of the pores in the solid phase, with formation of a certain number of vacant lattice points. These then disappear through interaction with dislocations, and not by elimination by way of the outer faces of the specimen, as supposed by Pines. These vacancies which may determine change of volume of the specimen during sintering are assumed to be of the Schottky type. Mathematical analysis gives an approximate theoretical expression for volume shrinkage during initial period of isothermal sintering of single-component metal powder bodies.

103. Rhines, F. N., *et al.*, "Rate of Densification in the Sintering of Uncompacted Metal Powders" in *Agglomeration*, ed. by W. A. Knepper, Interscience, New York (1962), pp. 351-369.

Investigations of the kinetics and energetics of densification carried out by simultaneous measurement of the variations of pore volume and surface area with time. These investigations showed that surface area decreases smoothly with time. In absence of chemical side effects, specific volume decreases smoothly with time. Also, it was found that internal surface area varies linearly with the density.

104. Ritzau, G., "Relationships between Thermoelectric Potential and Course of Sintering," *Planseeber. Pulvermet.* 8(3):100-109 (Dec. 1960).

In study of sintering process, obvious advantages attach to any method which permits succession of events to be followed during process itself, instead of periodically interrupting it in order to determine what changes have taken place during each successive stage. Author finds such a method in measurement of thermoelectric potential, i.e., the potential which arises in a closed circuit of two different metals when these are at two different temperatures at each of the two areas of contact.

105. Roberts, J. P., and J. Hutchings, "Nonstoichiometry of Zinc Oxide and Its Relation to Sintering," *Trans. Faraday Soc.* 55:1394-1399 (1959).

106. Roman, O. V., and H. H. Hausner, "Shrinkage of Copper and Iron Powder Compacts during Sintering," *Metal Progress* 83(2):104-108, 126, 128, 130 (Feb. 1963).

Four types of powders were studied to determine effect of factors such as particle size distribution, compacting pressure, compact dimensions, and sintering temperature and time; results indicate that changes in processing conditions can vary ratio of radial shrinkage to axial shrinkage from 0.5 to 3.

107. Roman, O. V., and H. H. Hausner, "Investigation in Linear Shrinkage of Metal Powder Compacts during Sintering," *J. Japan Soc. Powder Met.* 9(6):228-236 (Dec. 1962).

Study of variables which affect linear shrinkage of metal powder compacts during sintering; tests were made with several types of Cu and Fe powders as well as with Ni-Fe compositions, compacted and sintered under various conditions; it is shown that in changing these processing conditions, ratio of shrinkage in pressing direction to that of shrinkage perpendicular to this direction may change in value from approximately 0.5 to approximately 3; trends in these shrinkage ratio changes have been established for several variables in powder metallurgy processing.

108. Rudner, M. A., "Sintering of Polytetrafluoroethylene (Teflon)" in *Fluorocarbons*, Reinhold, New York (1958), pp. 58*ff*.

Sintering at 700-740°F in hot air furnaces. Rate of preheating up to the gel point (620°F) is critical. Sintering time is approximately 1 hr. Cooling rate affects dimensions and physical properties strongly.

109. Rutkowski, W., "Über den Einfluss des Sauerstoffgehaltes auf die Eigenschaften von Sintermetallen und die zwischen ihnen gebildete Kontaktzone," *Z. Metallk.* 51: 59-61 (1960).

Compacts of Cu and W powder and Fe and W powder, respectively, were produced in such a way that one half of the compact was W powder and other half was the other powder. The compacts with various amounts of oxides of the components added were sintered. The hardness in contact zone, electrical conductivity in neighborhood of contact zone, tensile strength of test samples, microstructure, and density of both parts of the samples were investigated. Tests show that amount of oxides affects the sintering mechanism. The optimum of properties is obtained with specific oxide amounts.

110. Rutkowski, W., "Studies on the Sintering Process and of Recrystallization Phenomena," *Neue Hütte* 3(1):37-43 (Jan. 1958).

Dependence of powder metallurgy on various other branches of science and technology pointed out. Powder metallurgy itself is a useful aid to the study of phenomena falling within the scope of other disciplines. Exemplified by reference to two investigations: (1) concerns the application of science to study of sintering process and influence of oxides in two selected systems; and (2) use of powder metallurgy as aid to study of recrystallization.

111. Samsonov, G. V., and M. S. Kovalchenko, "Some Features of the Sintering of Powders of Refractory Compounds," *Poroshkovaya Metallurgiya* 1(1):20-29 (Jan.-Feb. 1961).

Deals with both cold pressing and sintering and hot pressing of powders of the hard refractory carbides, nitrides, borides, and silicides, and compares their behavior with that of metal powders. Since the course of sintering is determined largely by characteristics of pressed compact, a preliminary discussion of the pressing properties of powder is necessary. Compaction found to be subject to same general laws as in case of metal powders, but the elastic aftereffect is more intense, and is dependent on the compacting pressure. In isothermal sintering, a state of constant density is reached which is explained by cessation of creep of particles. Both in sintering and in hot pressing, an initial stage is characterized by rapid shrinkage, which is associated with deformation of particles and collective recrystallization, and these lead to a stage of slowing down of the shrinkage process. Hot-pressed specimens, when the external pressure is removed, undergo some increase of dimensions, due to volume relaxation. Time of relaxation was found to decrease with rising temperature by an exponential law.

112. Schreiner, H., and F. Wendler, "Rekristallisation und Kornwachstum beim Sintervorgang," *Metall* 17(7):684-690 (July 1963); see also *Z. Metallk.* 52(4):218-228 (Apr. 1961).

Recrystallization and grain growth during sintering, particularly of sintered thermo-electric Peltier materials Bi_2Te_3 and $Sb_2Te_3-Bi_2Te_3$. Study of effects of sintering atmosphere, compacting pressure, sintering time, and sintering temperature on structure, which affects thermoelectric properties, showed that at temperature of 380°C, which results in greatest density, holding to 100 hr does not cause grain growth; increase of sintering temperature to 500°C and increased stress were without significant effect; vacuum sintering increased recrystallization but not grain growth.

113. Seigle, L. L., "Role of Grain Boundaries in Sintering" in *Kinetics of High-Temperature Processes*, ed. by W. D. Kingery, John Wiley, New York (1959), pp. 172-179.

    In porous compacts of Cu, Ni, and Al_2O_3, grain boundaries play an important role in densification. They are of negligible importance in sintering, which involves only interparticle bonding.

114. Seigle, L. L., and A. L. Pranatis, "Factors Affecting Sintering" in *Powder Metallurgy in Nuclear Engineering*, ASM, Cleveland (1958), pp. 10-30.

    At least four material transport processes operate during sintering of a powder compact: evaporation and condensation, surface diffusion, volume diffusion, and bulk flow. As particle size is decreased and ratio of surface area-to-volume increased, surface diffusion becomes relatively more important as a material transport process. Volume diffusion is the rate-controlling process in all but the initial stages of sintering of normally sized particles of nonvolatile metals and oxides.

115. Shmykov, A. A., and V. S. Saklinskii, "Effect of Allotropic Transformations on the Sintering of Iron Powder," *Metallovedenie i Termicheskaya Obrabotka Metallov*, Dec. 1960, No. 12, 26-30, 35-36 — Translated by H. Brutcher #5530.

    Dilatometric determination of linear shrinkage of pressed Fe powder samples between room temperature and 1000°C. Typical dilatometer record. Changes in shrinkage coefficient in various temperature ranges. Shrinkage above recrystallization threshold; final shrinkage after cooling and variation of total shrinkage as function of sintering temperature. Mechanical properties and electrical resistivity at various temperatures and holding times. Applicability of linear shrinkage coefficient within various temperature ranges and method of its determination to studies of the sintering of other metals and alloys. Procedure recommended for iron powder parts of improved strength.

116. Stone, F. S., "The Kinetics and Mechanisms of the Reactions of Solids" in *Reactivity of Solids: Proc. 4th Intern. Symposium*, Elsevier Amsterdam (1961), pp. 7-23.

    Discusses nature of reactions in solids, which necessarily involves consideration of mechanism of diffusion. Modern concepts of diffusion in solids place increasing emphasis on the part played by lattice imperfections. Theoretical work on point defects and dislocations has been followed by experimental investigation, which has shown basic ideas on imperfections in crystals to be very soundly based. Application of this knowledge in field of metal oxidation led to detailed interpretation of the parabolic law. Developments in this field are summarized. Mechanisms operative in sintering are discussed in light of some of the most significant publications on this subject. Reference is made to three stages of sintering: smoothing of particle surfaces, adhesion or welding together of particles, and elimination of voids. Value of experimental studies on systems of defined geometry pointed out. Attention drawn to part played by dopants, sintering atmosphere, and nonstoichiometry. Final sections concerned with thermal decomposition reactions and reactions involving release of trapped atoms, as in the annealing treatment of radiation damage.

117. Thümmler, F., "Wissenschaftliche Grundlagen der Sintervorgänge" in *Fortschritte der Pulvermetallurgie I*, by F. Eisenkolb, Akad. Verlag, Berlin (1963), pp. 329-452.

    Complete review of theories of sintering. Definitions. Sintering of single and multicomponent systems. Primary and secondary factors which affect sintering. Mechanisms of material transport. Methods of following up sintering. 323 references.

118. Tikkanen, M. H., "Eine optische Methode für kontinuierliche Schrumpfungsmessungen während des Sinterns," *Jernkontorets Ann.* 147(1):52-56 (1963).
      Optical method for continuous shrinkage measurements during sintering method described, successfully applied in Finland since 1955, makes it possible to record accurately changes of samples at very short time intervals during high-temperature sintering; its application to sintering of TiC–Ni is reported.

119. Tikkanen, M. H., "The Part of Volume and Grain Boundary Diffusion in the Sintering of One-Phase Metallic Systems," *Planseeber. Pulvermet.* 11(2):70-81 (Aug. 1963).
      Investigation undertaken to show possibility of measuring activation energies in sintering of metal powders more accurately than is usually done and further to show that both the volume and grain boundary diffusion processes are working during sintering process. Powders used were carbonyl Ni and atomized Ni powder. Shrinkage curves determined by following equation: $a/(1 - a) = (k_0 t)^n$, where k_0 is the rate constant, n is a constant, and $a = (V_0 - V_s)/(V_0 - V_{te})$, V_0 is the volume of unsintered compact, V_s is the volume of sintered compact at time t, V_{te} = volume of compact at 100% density.

120. Tikkanen, M. H., *et al.*, "Sintering of Metal Powder Compacts Containing Ceramic Oxides," *Powder Metallurgy* (1962) No. 10:49-60.
      Dimensional change during sintering of Co and Ni powders containing oxide dispersions studied; presence of MgO or CaO in Co and Ni retards densification of metals; retarding influence has maximum for certain proportion of oxide in metal; wettability between dispersed phase and metal may explain differences in sintering behavior.

121. Torkar, K., and H. Neuhold, "Surface Diffusion in the Sintering of Uncompacted Powders," *Z. Metallk.* 52:209-214 (1961).
      Authors employed magnetic method to enable them to separate surface diffusion from other types. Unpressed Cu–Ni powder mixtures were used, and method depends on fact that these two metals form a continuous series of solid solutions which remain ferromagnetic at room temperature up to Cu content of 32 w/o. The boundary line between ferromagnetic and nonferromagnetic zone can be made visible in microscope by means of a magnetic suspension. Measurements made indicate that surface diffusion plays an important role up to high temperatures in sintering of unpressed powders.

122. Vacek, J., "Über die Beeinflussung des Sinterverhaltens von Wolfram," *Planseeber. Pulvermet.* 7:6-17 (1959).
      Effect of metal additions on sintering behavior of W powder. Ni and Co, more than Fe, increase reactivity during sintering. Other metals do not show any marked influence. The influence of Ni was rather closely studied: amount, conditions for reduction of original W powder, pressure applied in compacting, sintering temperature, and time being varied widely.
      Influence of Ni on sintering of W powder studied by measuring density, hardness, specific electrical resistivity and transverse rupture strength of as-sintered specimens. Ni additions of as little as 0.25% (sintered 1 hr at 1300°C) resulted in almost pore-free W samples. Since this action can only be attributed to metals forming solid solutions with W, the formation of vacancies during diffusion is considered the primary reason for it. Diffusion examinations with the Fe metals have shown that Ni is most rapidly dissolved in W.

123. Vacek, J., "Relationship between Properties of Tantalum and Niobium Powders and Their Sintering Behavior," *Powder Metallurgy* (1961) No. 7:156-166.

Examination of effect of C and Cb on hardness and tensile and electrical properties of semimanufactured products of tantalum; influence of particle size distribution of Cb powder upon properties of sintered bars also studied, with particular reference to problem of swelling of bars during sintering.

124. Van Bueren, H. G., and J. Hornstra, "Grain Boundaries and the Sintering Mechanism" in *Reactivity of Solids: Proc. 4th Intern. Symposium,* Elsevier, Amsterdam (1960), pp. 112-121.

Absorption of vacancies at grain boundaries, which is considered as one of the fundamental mechanisms of sintering, is investigated in detail. Function of a particular grain boundary as a sink for vacancies is intimately connected with its structure and mobile properties. A twist boundary cannot absorb, a symmetric tilt boundary can, if the two grains on both sides of it move toward each other; whereas the boundary itself remains stationary. Considerably more effective sinks are asymmetric grain boundaries which can be induced to absorb vacancies by letting the two grains slide along each other, the boundary itself assuming a curved shape. The efficiency ratio for vacancy absorption between a "shearing" and a "squeezing" boundary may amount to a factor 6; moreover, in first case the shearing motion aids in the closing of pores, the sources of excess vacancies. The sintering is, therefore, to be considered as a combination of diffusion and plastic flow. The described mechanism leads to reasonable sintering times proportional to the pore volume, and to a temperature dependence according to that of self-diffusion.

125. Vasilos, T., and R. M. Spriggs, "Pressure Sintering—Mechanisms and Microstructures for Alumina and Magnesia," *J. Am. Ceram. Soc.* 46(10):493-496 (Oct. 1963).

Pressure sintering of pure alumina and magnesia in graphite dies and in ceramic dies over range 1100-1700°C and 4000-10,000 psi is useful in fabricating these materials with controlled microstructures, e.g., very high relative density and fine grain size or controlled larger grain sizes, without use of additives to control grain size. Apparent diffusion coefficients calculated for densification process are orders of magnitude greater than for pressureless sintering, which might be explained by enhanced diffusion under stress. Alternatively, lower calculated coefficients, which are more in agreement with pressureless sintering data, result when pressure correction terms modified by porosity are applied to existing relations.

126. Watanabe, T., "Mechanism of Pore Formation During Sintering of Fe–Cu Porous Compacts," *J. Japan Soc. Powder Met.* 8(2):63-72 (1961) (in Japanese).

Sintering of Fe-Cu compacts studied with view to their use as self-lubricating bearings. At temperatures above the $\alpha-\gamma$ transformation point of Fe, existing pores in compact found to grow slightly, owing to shrinkage accompanying phase change. At temperatures above melting point of Cu, the molten Cu spread into the pores, leaving new "copper-off" pores. Size of both old and new pores was somewhat enlarged by erosive action of molten Cu, and number of pores increased as result of expansion of compact accompanying Fe–Cu alloy formation.

127. Weise, G., *et al,* "Kornwachstumshemmung und Legierungsspuren in Karbonyleisen beim Sintern," *Die Bergakademie* 10:316-321 (1958).

128. Westerman, E. J., "Sintering of Nickel-Base Superalloys," General Electric Report 61-RL-2664-M (Mar. 1961).

Prealloyed powders of Ni-base superalloys were sintered to almost theoretical density in short sintering time. It was ascertained that rapidity of densification was caused by presence of small amount of liquid phase at grain boundaries, apparently originating by incipient melting in these regions. It was determined experimentally that this liquid phase formed at temperatures considerably below those at which macroscopic melting of compacts occurred. An increase in quantity of liquid phase present during sintering caused by an increase in sintering temperature, increased the densification rate, but resulted in a decrease in room-temperature tensile properties. Optimum properties were obtained by sintering for long times with a relatively small amount of liquid phase present.

129. Whalen, T. J., and M. Humenik, "Mechanisms and Microstructural Aspects of Liquid-Phase Sintering" in *Progress in Powder Metallurgy 1962, Proc. 18th Annual Technical Conf.,* Metal Powder Ind. Federation, pp. 85-98.

Considerable differences observed in liquid-phase sintering behavior of different compositions. Examples given to show that high rate of densification, grain growth, and rounding of the grain shape are not inherent characteristics of the process. It is, of course, necessary that the liquid phase be retained within the compact. Subject of wetting discussed. Processes taking place during actual sintering with formation of liquid phase are discussed under "particle rearrangement," "solution-reprecipitation," and "coalescence."

130. White, J., "Basic Phenomena in Sintering" in *Science of Ceramics,* ed. by G. H. Stewart, Brit. Ceramic Soc. and Neder. Keram. Vereniging (1961) Vol. 1, pp. 1-19.

Review of work of Ya. I. Frenkel (1945) to most recent publications on subject. Volume-diffusion model of Kuczynski, Kingery, and Berg, Coble *et al.,* and influence of grain growth in sintering as studied by Burke and Coble, are discussed, with particular reference to sintering of oxides. Some results by author and co-workers are quoted and interpreted in terms of relationship between surface energies of the system, and equations given which express conditions for equilibrium.

131. Whitmore, D. H., and T. Kawai, "Kinetics of Initial Sintering of Vacuum Reduced Titanium Dioxide," *J. Am. Ceram. Soc.* 45(8):375-379 (Aug. 1962).

TiO_2 in form of single-crystal plates and spheres. Measurements made of diameter of necks formed during heating in vacuum in temperature range 1200-1275°C. Comparison of curves obtained with sintering equations by Kuczynski showed that data corresponded to a fifth-power law, i.e., the predominant material transport mechanism was volume diffusion. Oxygen self-diffusion coefficients calculated from data agreed with those using rutile single crystals and an oxygen isotopic exchange technique. It is therefore considered likely that the rate-controlling step in the sintering process in question was oxygen ion diffusion.

132. Wilder, D. R., "Progress in Sintering," *Electrochem. Tech.* 1(5/6):172-178 (May-June 1963).

Review of representative items illustrating evolution of present concept of sintering, of which three stages can be distinguished: initiation, densification of material transport, and recrystallization and grain growth. Sintering is continuous process and the three stages are often interwoven.

133. Williams, J., *et al.* "The Effect of Surface Characteristics on the Sintering of Uranium and Beryllium Powders," *Powder Metallurgy* (1961) No. 7:167-188.

Studies of sintering of uncompacted U powders in vacuum have shown that sintering behavior is markedly affected by nature of surface films on powder. Of variables in powder production route that might affect the nature of surface films, the leaching stage is most important. Deleterious effect of surface contaminants on vacuum sintering behavior of Be is demonstrated. During development of a technique for the production of stable powder from electrolytic flake, nature of leaching treatment again proved to have a marked influence on sintering behavior of resultant powder.

134. Yao, T., and Y. Anekoji, "Sintering of Tungsten Metal Powder," *J. Japan Soc. Powder Met.* 9(5):163-168 (Oct. 1962) (in Japanese–English abstract).

Influence of particle size distribution, specific surface area, mixing ratio of fine and coarse powder, and compacting pressure on rate of sintering of W metal powder studied; it was found that rate of sintering (sintered density-green density/true density-green density) is proportional to specific surface area and is affected by particle size distribution of powder having same average particle size, but not affected by packing density.

135.   Yao, T., "Initial Sintering of Tungsten: Relation between Linear Shrinkage and Particle Size, Sintering Temperature and Time," *J. Japan Soc. Powder Met.* 9(6): 217-221 (Dec. 1962).

Compacts consist of various mixtures of coarse, medium, and fine powders; relations between linear shrinkage, particle size, sintering time, etc. are in good agreement with model of "bulk diffusion" proposed by Kingery, Berg, and Coble; shrinkage of triple mixture compacts are well explained by model; activation energy for sintering process is about 86,000 cal/mole.

136.   Youssef, H., and M. Eudier, "Frittage en phase solids de poudre de métaux purs," *Mem. Scient. de la Rev. Met.* 60(3):165-170 (Mar. 1963).

Solid state sintering of unalloyed metal powders; on basis of previous work, literature, further experimental study and theory, four mechanisms involved in sintering at different stages are discussed; these include some practical effects, such as roles of heating rate and of differences in specimen volume in shrinkage; stresses producing sintering could be calculated as function of surface tension.

137.   Zino, A. J., *et al.,* "Vacuum Sintering of Metal Powder Parts," *Metal Progress* 83 (4):88-92 (Apr. 1963).

High operating temperatures possible with vacuum sintering furnaces, up to 2300°F, can be utilized to increase density of metal powder parts, resulting in stronger product; process of sintering high-strength structural parts is now made more economical due to recent improvements in design and construction of vacuum heat-treating furnaces; encouraging test results on type 316 stainless, Monel, Fe–C, and Fe–C–Cu were obtained with standard MPIF tensile bars briquetted according to MPIF standards and sintered in vacuum at temperatures beyond those commercially practical today.

INDEX